Marcos Goes Oliveira

Genetic diversity of acerola trees (Malphigia emarginata D.C.)

Marcos Goes Oliveira

Genetic diversity of acerola trees (Malphigia emarginata D.C.)

Genetic diversity using molecular markers and morpho-agronomic characteristics in chard trees.

ScienciaScripts

Imprint

Cover image: www.ingimage.com

This book is a translation from the original published under ISBN 978-3-330-76392-0.

Publisher:
Sciencia Scripts
is a trademark of
Dodo Books Indian Ocean Ltd. and OmniScriptum S.R.L publishing group

120 High Road, East Finchley, London, N2 9ED, United Kingdom
Str. Armeneasca 28/1, office 1, Chisinau MD-2012, Republic of Moldova, Europe
Managing Directors: Ieva Konstantinova, Victoria Ursu
info@omniscriptum.com

Printed at: see last page
ISBN: 978-620-8-40798-8

To my dear aunt Judite, dear Camilla and my marvellous family, who have always supported me in my choices and, without their support and dedication, I wouldn't have made it to postgraduate studies.

I dedicate and offer this work.

ACKNOWLEDGEMENTS

To God, for the immense blessings he has conceived, and to my parents and all my family for their encouragement.

To UENF and CCTA for the opportunity to study for a master's degree.

To FAPERJ, for the grant.

To Professor Jurandi Gonçalves de Oliveira, for his teachings, encouragement, friendship and guidance.

Researcher Guilherme Eugênio Machado Lopes, from Pesagro, and friend Aroldo Gomes Filho, for their support during the experiment.

To Professors Messias Gonzaga, Alexandre Pio Viana and Gonçalo Apolinário, for their contributions and suggestions.

To the technician at the DNA marker laboratory, Vitória, for her help and dedication.

To my great friends from my hometown, especially Luciana, Charles, Grazi, Marilene, Francinaide and Roberta.

To my companions and friends from the Republic, Sérgio, Pedro, Sávio, Francisco, Ricardo and Adriano.

To the friends I made at UENF, who directly or indirectly contributed in some way.

SUMMARY

SUMMARY

Oliveira, M. G.; M. Sc., Universidade Estadual do Norte Fluminense Darcy Ribeiro, April 2008; Genetic diversity through morpho-agronomic characteristics and RAPD markers in acerola (Malpighia emarginata D.C.). Supervisor: Prof Jurandi Gonçalves de Oliveira.

Knowledge of the genetic variability and relationship between different acerola accessions is important in order to maximise the use of genetic resources for future breeding programmes. The aim of this study was to assess the genetic divergence between acerola accessions using morpho-agronomic characteristics and RAPD molecular markers. We used 48 accessions from an acerole population sown at the Pesagro - RIO experimental station, located in the municipality of Itaocara-RJ. The work was divided into two stages: morpho-agronomic characterisation, based on the list of minimum descriptors for this species; physico-chemical evaluations and molecular diversity using RAPD markers. For the morpho-agronomic evaluations, statistical analyses were carried out using principal components and the relative importance of the characters. For the RAPD, the distance matrix was used, based on the arithmetic complement of the Jaccard index, to group the accessions using the Tocher Optimisation and UPGMA methods. The results showed that there was genetic variability between the accessions studied. Thus, three quantitative characteristics considered to be of greater relative importance for the study of genetic divergence were selected: stem diameter, mature leaf length and maximum leaf width. For the physico-chemical evaluations, there was an increase in pH and TSS between the **accessions from the 'once' stage to** the semi-mature stage. Fruits **in the 'once' stage** showed higher values of ascorbic acid content, regardless of the access, with a gradual reduction until the final ripeness of the fruit, indicating that there is a possible correlation between ascorbic acid content and the stage of ripeness. The RAPD marker technique was effective for studying diversity among the accessions, showing that the population has ample genetic variability. The results showed that the accessions ACE 001, ACE 014, ACE 031, ACE 038, ACE 043 and ACE 046 can be indicated for vegetative propagation and clonal evaluation, as they had a high ascorbic acid content, reddish ripe fruit skin colour and wider and taller fruit, which defines the shape as globular.

CHAPTER 1

INTRODUCTION

The acerola tree (Malpighia emarginata D.C.) originated in the West Indies and later spread to other regions of the world, establishing itself particularly in tropical and subtropical ecosystems on the American continent (De Rosso & Mercadante, 2005). In Brazil, the acerola has been known for over 60 years (Neto et al., 1995). However, the cultivation of acerola has grown rapidly in the last 20 years and, currently, with the increase in demand for natural foods, acerola has seen a big boost in consumption (Moura et al., 2007).

According to Matsuura et al. (2003), acerola is nutritionally important due to its high ascorbic acid content and is a product widely used in the pharmaceutical and food industries. However, according to Assis et al. (2001), in addition to the high ascorbic acid content, acerola also contains significant amounts of vitamin A, iron, calcium, phosphorus, riboflavin and niacin.

In Brazilian orchards, there is a high degree of variability between cultivated genotypes with regard to important characteristics such as productivity, growth habit and plant size, canopy architecture, fruit colour, flavour, consistency and size, as well as pulp yield, among others, which is attributed to seminal propagation (Paiva et al., 1999). This high variability of genotypes has caused serious problems for the production system, as it makes it difficult to carry out all the cultural practices rationally, disorganising the producer's marketing system.

On the other hand, the occurrence of this variability among the genotypes grown in Brazilian orchards can be exploited in plant breeding programmes by selecting superior individuals, based on characteristics of interest to the consumer market, as well as genotypes that are better adapted to the country's various producing regions.

In Brazil, some research institutions have been working on launching new materials, providing new prospects for producers. These include Flor Branca, Inada, Número 1, Número 54/02, Okinawa and Sertaneja. Soares Filho & Oliveira (2002) report that the only cultivar recommended for commercial planting is Sertaneja Brs, which was launched by Embrapa Semi-Arid (CPATSA). In commercial plantations, production varies from 20 to 50 kg of fruit/plant/year (Alves et al., 1995).

The main barrier encountered in acerola breeding programmes is obtaining cultivars with higher productivity and vitamin C content. The main aim of these crop improvement programmes is to select superior genotypes, or to generate clones or populations with greater genetic uniformity, providing fruit that satisfies the most different consumers, in order to conquer markets throughout the country and the world (Paiva et al., 1999).

Currently, genetic improvement programmes have used the combination of classical techniques with biotechnological tools, such as the use of molecular markers, with substantial gains in terms of reducing the time taken to identify genetic diversity among the individuals being worked on (Xavier et al., 2005).

Molecular markers such as RAPD (polymorphic amplification of arbitrary DNA), because they are a fast, relatively low-cost technique with informative potential (Williams et al., 1990), have been used to study genetic diversity in acerole (Salla et al., 2002), banana (Souza, 2006), açaizeiro (Oliveira et al., 2007) and passion fruit (Viana et al., 2003).

The aim of this study was to assess the genetic diversity between accessions of an acerole population using RAPD molecular markers and morpho-agronomic descriptors, with the aim of identifying promising genotypes for use in crop improvement programmes.

CHAPTER 2

LITERATURE REVIEW

2.1 Economic importance

The acerola (Malpighia emarginata D.C.) has attracted the interest of fruit growers and has become economically important in various regions of Brazil, due to its potential as a natural source of vitamin C and its great capacity for industrial utilisation (Nogueira et al., 2002).

Brazil stands out as the world's largest producer, consumer and exporter of acerola, with an area planted in the Northeast region of more than 2,000 ha (França & Narain, 2003). There are commercial acerola plantations in practically every Brazilian state (Alves, 1996). According to Donadio et al. (1998), acerola is grown on a larger commercial scale in the states of Bahia, Pernambuco, Paraíba, São Paulo, Paraná, Rio Grande do Norte, Pará and Amazonia. However, it is in the Northeast region that, due to its favourable soil and climate conditions, the acerola tree has adapted best (Paiva et al., 1999).

Most of Brazil's acerola production is linked to the agro-industrial sector (Coelho et al., 2003), with a view to utilising the fruit. However, a considerable part of this production is not utilised due to the high perishability of the fruit, which means that around 60% remains in the domestic market and 40% is sent abroad (Oliveira & Soares Filho, 1998), mainly to Japan, Europe and the United States (Coelho et al., 2003).

Acerola has potential for industrialisation, as it can be consumed in the form of jams, jellies, used to enrich juices and diet foods, in the form of nutraceutical foods such as tablets or capsules, used as food supplements, teas, sports drinks, cereal bars and yoghurts (Carpentieri-Pípolo et al., 2002). In addition, acerola can be consumed in the form of whole, concentrated or freeze-dried juice, liqueur, chocolates, chewing gum, nectars, puree, ice cream, biscuit toppings, soft drinks, among others (Carvalho, 2000). However, the most common forms of acerola commercialisation are fresh fruit, frozen pulp and bottled juice (Yamashita et al., 2003).

1.1 Centre of Origin and Introduction of Aceroleira in Brazil

The acerola tree or "Antillean cherry" *(Malpighia emarginata* D.C.) is a plant typical of countries with a tropical climate, with its centre of origin in the Antillean Sea region, northern South America and Central America (Lopes & Paiva, 2002).

For a long time, this "tropical cherry tree" remained in bloom and fruiting on American soil without attracting much attention. It is believed that the cultivation of this plant began to gain momentum in 1946, when Asenso and Guzman, quoted by Marty and Pennock (1965), discovered the high vitamin C content of its fruit. Commercial planting of the acerole tree began in Puerto Rico and spread to Cuba, Florida, Hawaii and other parts of the world.

In Brazil, the acerola tree was introduced in 1955 in the north-east through the Federal Rural University of Pernambuco, with seeds brought from Puerto Rico (Simão, 1971). However, according to Andrade et al. (1995), acerola cultivation only became commercial in the 1980s, with the pioneering states of Bahia and Pará aiming to export acerola to Europe and Japan.

2.2 Botanical Aspects and Cytogenetics

The botanical classification of acerola has been controversial. Despite the adoption of the name *Malpighia emarginata* by the International Plant Genetic Resources Council in 1986, this name is still rarely used, as it was initially classified as M. punicifolia and M. glabra. However, Asenjo (1980) reports that the names M. glabra and M. punicifolia are synonyms, but applied to a different species of acerola tree, the species being called Malpighia emarginata D.C.. Nogueira (1997), who was studying the physiological expressions of the acerola tree, sent several samples to Dr Willian R. Anderson, a specialist in the Malpighiaceae family, who found that it was a single species, M. emarginata D.C., which is correctly used to refer to the acerola tree.

The acerola tree is a Magnoliopsida belonging to the Malpighiaceaes family, which is made up of 39 to 41 species of shrubs and small trees. According to studies cited by Alves & Menezes (1995), the acerola tree is classified as follows:

Division: Tracheophita

Subdivision: Spermatophitina

Class: Magnoliopsida

Family: Malpighiaceae

Genus: Malphigia

Species: Malphigia ermaginata D. C.

In order to obtain information on the number of chromosomes and pollen fertility, Singhal et al. (1985) studied some species of Malphigeaceae and found that they had a chromosome number of 2n = 20 and 90% pollen grain fertility. Cavalcante et al. (1998), in cytogenetic studies carried out on eight acerola genotypes selected by Embrapa Mandioca e Fruticultura, observed that all the genotypes studied were diploid with 2 n= 20 chromosomes.

2.3 Description and Characteristics of the Acerole Tree

Ledin (1958), quoted by Araújo & Minami (1994), describes the acerole tree as a medium-sized glabrous shrub, 2 to 3 metres high, with a crown diameter of up to 3 metres. The plant also has a single stem or small trunk, often branched, with a dense crown made up of numerous scattered woody branches, usually curving downwards. The leaves are elliptical, oval, 2 to 7.5 cm long and 1 to 6 cm wide.

According to Simão (1971), the fruit of the aceroleira is a rounded drupe, with a diameter of between 1 and 3 cm and a weight of between 3 and 16 grams. The size of the fruit varies depending on the genetic potential of the plant, cultivation and the number of fruits per axil. The colour of the ripe fruit can vary from red, purple or yellow.

The acerola flower has some of the characteristics of an autogamous species: the complete flower with androecium and gynoecium, simultaneous ripening of the androecium and gynoecium and the absence of physical barriers that prevent self-fecundation. Associated with these characteristics, it is common for fruit to be produced on single plants. However, Lopes et al. (2002) in a study estimating the crossing rate of the acerole tree, based on isoenzyme data, concluded that the acerole tree is a predominantly allogamous species, due to the great phenotypic variability observed in the orchards, which suggests the occurrence of recombination. According to these authors, self-fertilisation of flower buds under experimental conditions promotes less fruit set when compared to manual crossing and natural pollination.

The fruit develops very quickly, taking 22 days from flowering to ripening (Batista et al., 1991; Gomes et al., 2001). Knowledge of the length of the crop cycle is important for acerola growers, who can then programme their harvesting activities and commercialise the fruit on the domestic and foreign markets with a greater chance of success.

The acerole tree can produce four to six blooms a year if well managed (Almeida & Araújo, 1992). This response is basically due to the soil and climate conditions associated with the practice of irrigation, which favour several growth spurts, thus providing almost continuous flowering and fruiting. It is important that the plant is adequately supplied, especially with nitrogen and water after flowering, as it is common for flowers subjected to adverse conditions to abort.

2.4 Genetic Improvement of the Acerole Tree

Generating new, more productive cultivars with superior quality characteristics such as fruit colour, taste, smell, texture and colour of the flesh, sugar content, acidity, resistance to transport, among others, has been the great challenge of genetic improvement of the acerola tree. According to Paiva et al. (1999), plant breeding aims to exploit existing genetic variability in order to obtain clones or populations with greater genetic uniformity, which produce fruit that satisfies the most diverse markets in the most developed and economically prosperous regions of Brazil and the world.

Several isolated studies have been carried out in some institutions, where basic studies are being developed to support the improvement of the acerola tree, such as those related to reproductive biology (Magalhães & Ohashi, 1997; Gomes et al., 1998), cytogenetics (Cavalcante et al., 1998), the genetic control of fruit characteristics (Lopes, 1999), among others. However, the most common procedures adopted for improving acerola have been the introduction of germplasm, clonal selection and hybridisation followed by selection.

The main methodology adopted in acerola breeding programmes is the selection of clones, because not only can the results be seen in the short term, due to the fact that acerola is a species that can be propagated vegetatively, but it is also the most efficient way to meet the immediate demand for varieties, where the genotype of each plant can be passed on in its entirety through the generations (Paiva et al., 2003).

Paiva et al. (2003), evaluating acerola germplasm with the aim of verifying the performance of 45 clones and selecting those that stood out in terms of plant growth, production and fruit quality, found four superior genotypes based on morphological characteristics. The selection was based on a combination of the characteristics of plant crown conformation, fruit production and vitamin C content, among others.

Bosco et al. (1994) selected nine acerola clones based on the phenological characteristics of the plant and the morphology of the fruit. Fruit weight ranged from 7.02 to 9.68 g; average fruit diameter was 2.51 cm and, in general, the selected clones had a consistency, colour and flavour that fully met market requirements.

In almost all the orchards set up in Brazil, there is a very marked mixture of types and forms of acerola trees, and it is common to find plants with different growth habits and different fruit shapes, colours and sizes in the same orchard. As a result, breeders have noticed that there is a need for research to characterise, select and disseminate genetic materials. Thus, each institution has adopted names and codes in the germplasm banks to identify and differentiate the fruit in relation to their respective genotypes.

2.5 Molecular Markers

A molecular marker can be defined as any phenotype resulting from the expression of a gene, as in the case of proteins, morphological characters, or a specific segment of DNA (corresponding to expressed or unexpressed regions of the genome), whose sequence and function may or may not be known (Ferreira & Grattapaglia, 1998). According to Milach (1998), a molecular marker is a genetically inherited DNA characteristic that differentiates two or more individuals.

Molecular markers are an important tool for genetic characterisation, thus helping to alleviate and rectify problems related to duplications in germplasm banks, as well as identifying divergent cultivars (Nass, 2001). Molecular markers are also used extensively and successfully in the genetic analysis of plants and in characterising the variability that exists between individuals.

Genetic analysis of individuals has evolved considerably in recent years, thanks to the development of the DNA sequence amplification technique, which allows direct comparison of alleles through the composition of the nucleotide sequence. This methodology, called PCR (Polymerase Chain Reaction), makes it possible to compare organisms at a molecular level, without the influence of the environment or age of the tissue (Ferreira & Grattapaglia, 1998).

There are various techniques that make it possible to directly identify DNA polymorphism and therefore mark specific sequences or parts of it. Markers such as RAPD, Microsatellite and AFLP are some of the most widely used in genetic studies Ferreira & Grattapaglia (1988).

Variations of PCR, such as RAPD (random amplifield polymorphic DNA), have been widely used for genetic studies in fruit trees, including the use of molecular markers to analyse genetic variability in acerola (Salla et al., 2002); studies to characterise grape varieties (Ulanovsky et al., 2002); assessment of genetic diversity in banana (Souza, 2006) and coconut (Wadt, 1997). The RAPD technique is based on the cyclic repetition of the enzymatic extension of primers (small complementary DNA sequences), which are ringed at the two opposite ends of a DNA strand, which serves as a template. In this technique, only a single primer is used instead of a pair, as in PCR, and this primer has an arbitrary sequence and therefore its target sequence is unknown.

The RAPD technique, among those mentioned, is the one with the lowest cost, number of steps and time to obtain the results, and is the easiest to implement. However, it has the disadvantage of low repeatability and little consistency from one laboratory to another, which makes it difficult to compare data obtained in different locations. Care must therefore be taken to standardise the technique in the laboratory for characterising cultivars. The level of polymorphism obtained with RAPDs varies greatly according to the species in question and has been used successfully to characterise plants.

2.6 Morphological and Genetic Divergence

In plant breeding programmes, information on genetic diversity and divergence within a species is essential for the rational use of genetic resources (Loarce et al., 1996). Studies on genetic diversity in germplasm collections can be carried out using morphological characters of a qualitative or quantitative nature (Moreira et al., 1994).

Molecular characterisation of the genetic diversity of germplasm can provide useful data to help breeders identify and select the basic genitors or clones for establishing a breeding programme. Information on genetic diversity can help with breeding programmes, as well as avoiding redundancies or mixtures of genotypes in germplasm conservation studies and programmes.

According to Shimoya et al. (2002), genetic divergence is assessed using various methods that take into account agronomic, physiological, genetic and morphological characteristics. This provides information for identifying genitors that, when crossed, increase the chances of producing superior genotypes in segregating generations.

Various methods can be used to study genetic divergence. This choice is based on the precision desired by the researcher, the ease of analysis and the way in which the data was obtained. Multivariate analysis techniques can be used to assess the divergence between accessions and to select the most important descriptors for discriminating between accessions in a germplasm bank (Amaral Júnior, 1994).

Pípolo et al. (2000), in a study of multivariate genetic divergence in acerola, managed to divide 14 genotypes into three groups using the Tocher clustering method, based on the most important fruit quality character for this crop, vitamin C content. These authors highlight some promising crosses among these accessions.

Other genetic diversity studies have been carried out using molecular marker techniques and have proved to be very promising. Viana et al. (2003) studied the genetic diversity between commercial genotypes of yellow passion fruit (Passiflora edulis f. flavicarpa) and between species of native passion fruit, determined by RAPD markers. The authors carried out a cluster analysis between the genotypes using the complement of the Jaccard index and also used Ward's hierarchical clustering method, based on the matrix of genetic distances obtained by the complement of the Jaccard index. The species studied showed great genetic variability.

Ruas et al. (1995) cited the importance of estimating the genetic relationship and diversity of pineapple cultivars for the evaluation of genetic resources. These authors found a very similar genetic relationship between four pineapple cultivars (Pérola, Smooth Cayenne, Primavera and Perolera) obtained from analysing the information generated by means of RAPD molecular markers and analyses of morphological and agronomic characteristics. Thus, the combination of molecular analyses with morpho-agronomic evaluations can be used efficiently to characterise genetic resources in the Ananas genus.

Costa (2004) characterised the grape germplasm collection at UENF, assessing the genetic divergence between 40 genotypes of species, varieties, rootstocks and hybrids, using RAPD molecular markers. To do this, the author used multivariate statistical techniques, such as the Tocher clustering method and the hierarchical clustering method (UPGMA), which were in agreement.

Salla et al. (2002), analysing 24 acerola accessions belonging to the active germplasm bank of the State University of Londrina, detected high polymorphism among the accessions. The results showed that despite the narrow genetic base common in acerola collections found in Brazil, genetic variability is relatively high.

2.7 Multivariate Analysis

Multivariate analysis for evaluating genotypes based on quantitative traits seems to be the ideal method, considering the existence of correlations between these traits. From this analysis, it is possible for the information from each character to be complemented and the analysis as a whole to be more accurate than univariate analyses.

The use of multivariate techniques to estimate genetic divergence has been routinely employed by plant breeders. Among these practices, the most commonly used are principal component analysis, when the data is obtained from experiments without repetitions; canonical variable analysis, when the data is obtained from experiments with repetitions and, finally, grouping methods, the application of which depends on the use of a previously estimated dissimilarity measure (Cruz & Regazzi, 1997), for example, the Euclidean distance can be estimated based on data without repetitions, as is the case with data from the active germplasm bank, making its application feasible (Carvalho et al., 2003).

The use of multivariate analysis in studies on genetic divergence has been the subject of extensive research and has been used to evaluate different germplasm collections with a view to selecting superior materials.

There are several examples in the literature of the use of multivariate techniques to analyse genetic divergence in crops, including those carried out with Capsicum (Sudré et al., 2005), popcorn (Miranda et al., 2003), cocoa (Dias & Kageyama, 1997), guaraná (Nascimento filho et al., 2001), cotton (Carvalho et al., 2003) and beans (Coimbra et al., 1999; Rodrigues et al., 2002).

Lopes et al. (2000), evaluating the genetic divergence of 112 acerola accessions from the UFV active germplasm bank, taking into account the characteristics of the fruit, found six accessions with favourable characteristics, which were indicated for clonal evaluation, as well as five for use in hybridisation.

CHAPTER 3

MATERIAL AND METHODS

3.1 Genetic Material

This study used 48 accessions (ACE 001 - ACE 048) from a population of acerole trees approximately 15 years old, from the seedling production centre of the former CATI (Coordination of Integral Technical Assistance), located in São Bento do Sapucaí - SP, and sown at the Pesagro-RIO Experimental Station, located in the municipality of Itaocara-RJ. The municipality of Itaocara is located **in the north-west of Rio de Janeiro, at 21° 39' 12" S and 42° 04' 36" W, with an altitude of 60 metres, an AWi climate, with** an average annual temperature of 22.5°C and average annual rainfall of 1041 mm (Fontes, 2002).

3.2 Morpho-agronomic characterisation

3.2.1 Descriptors

The characterisation of Pesagro's acerola population was based on the list of minimum descriptors for this species (Oliveira et al., 1998). Due to the lack of initial fruiting of some of the accessions, for this study it was only possible to tabulate information on vegetative and inflorescence characters for 38 accessions. A total of 26 descriptors were used, 17 of which were qualitative and nine quantitative.

The leaf, flower and fruit descriptors are described below:

a) crown conformation (CC). Determined using a grading scale: 1-globular; 2-intermediate; or 3-Erect.

b) crown branching (RC). Determined using a grading scale, with: 1-little branched; 2-branched; 3-much branched.

c) leaf texture (FT). Scale of scores: 1-coreaceous or 2-pliable

d) leaf shape (FF). Determined using a grading scale, with: 1-rounded; 2-ovate; 3-elliptical.

e) general shape of the mature leaf edges (FGBF). Determined using a grading scale, with: 1- straight edge; 2- slightly wavy edge; 3- wavy edge.

f) general shape of the leaf limb (FL). Determined using a grading scale: 1- open; 2-intermediate; 3- closed.

g) presence of leaf pilosity (PF). Determined by the presence or absence of leaf hairiness.

h) presence of pilosity on the branch (PR). Determined by the presence or absence of pilosity on the branches.

i) flowering type (TFL). Determined using a grading scale and assigned: 1- single flowers; 2- presence of flowers in panicles; 3- both types of flowering.

j) colour of the flower corolla lobes (CLC). Determined using a grading scale: 1- white colour; 2- light pink colour; 3- dark pink colour; 4- violet colour.

l) uniformity of ripeness (UM). Determined by a grading scale 1- uniform; 2-uniform.

m) colour of the skin of the immature fruit (CCFI). Determined using a scale of scores: 1- green; 2- yellowish green; 3- purplish green; 4- greenish purple; 5- purple.

n) colour of the skin of the ripe fruit (CCFM). Determined using a scale of scores: 1- yellow; 2- orange; 3- pink; 4- red; 5- purplish red; 6- purple.

o) colour of the flesh of the ripe fruit (CPFM). The colour was determined using a scale of scores: 1- whitish; 2- green; 3- yellow; 4- pink; 5- red; 6- purple.

p) rind texture (TC). Determined using a grading scale: 1- smooth; 2- intermediate; 3- wrinkled.

q) fruit surface furrow (SSF). Determined using a grading scale: 1- superficial; 2- intermediate; 3- deep.

r) mature fruit size (MFS). Determined using a grading scale: 1- small fruit (12-16 mm); 2- medium fruit (17-22mm) or 3- large fruit (23-30mm).

The quantitative descriptors were as follows:

s) plant height (H): measured from ground level to the apical end of the highest branch using a tape measure, expressed in metres.

t) plant diameter (PD): measured at 100 cm from ground level using a tape measure, expressed in metres.

u) stem diameter (SD): measured at 10 cm from ground level using a tape measure, expressed in centimetres.

v) Mature leaf length (ML): measured from the base of the central vein of the median lobe to the end, expressed in centimetres, using the average of 20 leaves obtained from the median part of the canopy.

w) maximum leaf width (LF): made using the average of 20 leaves similar to the previous item, thickened in centimetres.

x) average number of flowers per panicle (N°MFP): each plant was sampled and counted in four quadrants.

y) average number of fruits per panicle (N°FP): each plant was sampled and counted in four quadrants.

3.2.2 Physical, Chemical and Physico-Chemical Assessment

Of the accessions studied, 30 per cent were vegetating. It was therefore necessary to select individuals with enough fruit for the chemical and physical analyses. Twenty-five accessions were selected and evaluated using fruit collected at three stages of ripeness.

The physical, chemical and physico-chemical characteristics are described below:

- Fruit diameter (Diam): average of 20 randomly selected fruits per access. The maximum diameter of each fruit was measured using a caliper and expressed in millimetres.

- Fruit length (Comp): average of 20 randomly selected fruits per access. The maximum length of each fruit was measured using a caliper and expressed in millimetres.

- Fruit colour (L*a*b*h°): colour was assessed by refractometry, with the average of 20 fruits, using the Minolta Chroma MeterCR310 digital device, which expresses it according to the CILAB system (L*a*b), according to Wolf et al. (1997).

- Soluble solids content (SS): average of 20 fruits per access chosen at random. The soluble solids content was measured using a refractometer on samples of juice extracted from the fruit (AOAC, 1970).

- Total titratable acidity (ATT): determined by neutralising the acidity of the fruit pulp by titration with 0.1N NaOH, according to the methodology described by A.O.A.C (1984).

- Vitamin C (Vit C) content: assessed from the macerated pulp by titration with 2,6 dichlorophenol-indophenol, according to the official A.O.A.C. method (1970), using about one gram of macerated fresh fruit and expressed in mg of ascorbic acid/100 g of pulp.

3.2.3 Statistical analysis

After the evaluations, the morphological characteristics were statistically analysed using computer resources and the Genes programme (Cruz, 2006). The qualitative descriptors were analysed using multicategorical variables, given that all the variables studied had more than two characteristics that were notably exclusive. To this end, Jaccard's coefficient was used as a measure of dissimilarity.

For the quantitative characters and physical and chemical evaluations, the principal components method was used, based on standardised data and the relative importance of the characters. The principal components technique was adopted because it is based only on the individual information of each access, without the need for repeated data (Cruz and Regazzi, 2001). 20 fruits were collected from each plant, from which an average of these individuals was obtained.

To represent the relationships between the accessions and characterise diversity, Euclidean distances were calculated on the variables, which allowed hierarchical dendrograms to be constructed using the UPGMA method and the Tocher optimisation method.

3.3 Molecular marker analysis

3.3.1 DNA extraction

DNA was extracted from young acerola leaves collected at the Pesagro - Rio experimental station. Branches 10-15 cm long were collected from the apex of the plant, wrapped in aluminium foil, placed in duly identified individual plastic bags and immediately placed in a Styrofoam container with crushed ice to maintain their integrity. This material was transported to the Molecular Markers department of the Plant Genetic Improvement Laboratory - CCTA/UENF. In the laboratory, the leaves used as samples were frozen in liquid N2 and macerated in a porcelain mortar, previously cooled to avoid raising the temperature and consequently denaturing the enzymes. The protocol adopted for extraction was adapted from the CTAB (Cethyltrimethyllammonium Bromide, Sigma) method in the extraction buffer described by Doyle & Doyle (1987).

Approximately 50mg of macerated tissue was transferred to 1.5ml tubes and immersed in liquid N2. 800|jL of pre-warmed extraction buffer was added to each tube, containing the solution prepared with 2% CTAB, 1.4M NaCl, 20mM EDTA, 100mM Tris-HCl (pH 8.0), 1% PVP and 0.2% B-mercaptoethanol, and then incubated at 65°C for 40 minutes, with gentle shaking of the tubes every 10 minutes. After this stage, the tubes were centrifuged at 14,000 RPM for 5 minutes. A fraction of the supernatant (800|jl) was transferred to new tubes, where 800|jL of chloroform: isoamyl alcohol (24:1) was added, followed by continuous inversion of the tubes until the solution became turgid. This step was repeated once more. After centrifuging again at 14,000 RPM for 5 minutes, the supernatant was transferred to new tubes, where two thirds of the volume of ice-cold isopropanol was added, followed by gentle inversion and then resting in an ultrafreezer at - 80 °C for 30 minutes. Centrifugation was again carried out at 14,000 RPM for 10 minutes, resulting in **a pellet. This** was washed twice with 300μl of 70% ethanol to remove any salt present, and once more with 300μl of 95% ethanol (between each wash, the material was centrifuged at 14000 RPM for 10 minutes). After discarding the last supernatant, the material was dehydrated under natural conditions for approximately 30 minutes. The precipitate was resuspended in 300μL of Tris-EDTA solution (10mM Tris-HCl, 1mM EDTA, pH 8.0) with RNAse at a final concentration of 40 μg/ml and incubated in a water bath at 37°C for 30 minutes. Next, 20 μL of 5M NaCl and 150 μL of ice-cold isopropanol were added to precipitate the DNA again, and this mixture was incubated at -80°C for 30 minutes. After this stage, the DNA was sedimented by centrifugation at 14,000 RPM for 10 minutes and washed twice with 70% ethanol and once with 95% ethanol. After dehydration, the final precipitate was suspended in 300µL of water.

3.3.2 DNA quantification

DNA quantification was carried out by visual comparative analysis on agarose gels subjected to electrophoresis (100V for 90 minutes). The gel was prepared with 0.8% (w/v) agarose, 0.5 X TAE (Tris base, sodium acetate, 0.5M EDTA and distilled water) and 4 µL of ethidium bromide (10 mg/ml). A solution of 2ML of DNA from each genotype, 3ML of Blue Juice dye and 5ML of TE solution were applied to the gel. The standard DNA, High DNA MASS Ladder (INVITROGEN), was used to compare the size of the fragments. After electrophoresis, the gel was exposed to ultraviolet light and immediately photographed using the Eagle Eye II equipment (Stratagene). The DNA was then diluted to the appropriate concentrations for use in the RAPD reactions.

3.3.3 Random amplified DNA polymorphism (RAPD) assay

After quantification, the DNA fragment amplification reactions were carried out in a Martercycler gradient thermal cycler (Eppendorf), in a final volume of 20 µL, containing: 10.80 µL of ultrapure water; 2 µL of amplification buffer (100 mM Tris-HCl pH 8.3, 50 mM KCl); 1.60 µL of MgC12 (25 mM); 1.0 µL of DNTPs (2 mM of each of the deoxyribonucleotides dATP, dTTP, dCTP, dGTP); 2.0 µL of primer (5 mM); 2.0 pL of genomic DNA (10 ng) and 0.6 unit of Taq DNA polymerase. 2.0 µL of DNA were added to the microtubes and, in parallel, a mix was prepared containing all the other reagents at the concentrations mentioned, where a different primer was added to each mix. From these solutions, 18 µL were removed and added to the microtubes, totalling 20 µL for the reaction.

A Perkin Elmer GeneAmp PCR System thermocycler was used, programmed for 45 amplification cycles after initial denaturation at 95 °C for one minute. Each individual cycle consisted of 1 minute at 95 °C, 1 minute at 36 °C and 2 minutes at 72 °C. After the 45 cycles, a final 7-minute extension step was carried out at 72 °C.

After the reactions, the amplified fragments were subjected to an electrophoresis run (100 v for 90 minutes) on a 2% (w/v) agarose gel, using 0.5 X TAE buffer. To compare the size of the amplified fragments, the 250 bp DNA Ladder purchased from INVITROGEN Life Technologies was used as a standard.

3.3.4 Selection of primers

In order to select the most informative primers, a screening was carried out using samples from four accessions at random, based on the list of minimum descriptors for acerola (Oliveira et al., 1998). Several Operon Technologies primers were tested and only the primers with clear bands and the highest number of polymorphisms among the genotypes tested were chosen. After selection, each access,

were individually tested with primers to assess DNA polymorphism, with the aim of selecting those that amplified simultaneously for all genotypes.

3.3.5 Statistical analysis of RAPD markers

The computational resources of the Genes programme (Cruz, 2006) were used to analyse the results.

The statistical analysis was based on the binary data matrix formed by analysing the RAPD gels, in which the presence of a band was assigned a value of 1, while the absence of a band was assigned a value of zero. The genetic distance was calculated in pairs between the genotypes by the arithmetic complement of the Jacard index, whose expression is: $C_{ij} = 1 - \frac{a}{a+b+c}$, where:

a= number of bands present in accessions i, j.

b= number of bands present in accession i and absent in accession j.

c= number of bands present in accession j and absent in accession i.

3.3.6 Grouping methods

Based on the resulting binary data table, an analysis of the genetic divergence of the 48 acerola accessions was carried out using the distance matrix, based on the arithmetic complement of the Jaccard index, which was used to group the genotypes using the Tocher Optimisation methods. This method adopts the criterion that the average of the dissimilarity measures within each group should be less than the average distances between any groups (Cruz, 2006). The UPGMA (Unweighted Pair-Group Method Using Arithmetic Average) hierarchical method was used to obtain the dendrogram.

CHAPTER 4

RESULTS AND DISCUSSION

4.1 Morpho-agronomic characterisation

4.1.1 Morpho-agronomic characterisation - qualitative characters

The data relating to the qualitative characteristics of the 38 accessions is shown in Table 1. The results obtained from analysing the multicategorical variables of the acerola accessions show that there is genetic variability in 70.5% of the characters studied. Of the 17 characters evaluated, there was no variability in five characteristics, namely leaf texture (TF), presence of leaf hairiness (PF), presence of branch hairiness (PR), type of flowering (TFL) and uniformity of ripening (UF). In all the accessions evaluated, the PR was malleable, no PR or PF was detected, the flowers showed isolated TFL and there was no UF between the accessions for fruit ripeness.

Table 1- Notes for the 12 qualitative descriptors in 38 acerola accessions. **CC** = crown conformation (1-globular; 2-intermediate; 3-straight); **RC** = crown branching (1-little branched; 2-branched; 3-branched); **FF**

Access	CC	RC	FF	FGBC	FL	CLC	CCFI	CCFM	CPFM	TC	SSF	TFM
ACE 001	3	3	3	3	1	2	1	6	5	1	1	2
ACE 002	3	1	3	3	2	2	1	5	3	3	3	2
ACE 004	1	1	3	1	1	3	3	4	3	1	1	2
ACE 005	1	3	3	2	2	3	1	5	4	1	1	3
ACE 007	3	3	2	3	1	2	1	5	5	3	1	2
ACE 008	1	3	3	3	2	3	1	5	5	1	3	3
ACE 009	2	3	2	3	2	1	1	4	3	1	2	3
ACE 010	3	3	2	3	1	1	1	4	3	1	1	2
ACE 011	3	2	3	2	1	2	1	4	3	1	2	2
ACE 012	3	3	3	3	2	3	1	4	3	3	2	2
ACE 014	3	3	3	3	2	2	1	4	3	3	2	2
ACE 016	2	3	3	3	1	3	1	6	5	1	1	3
ACE 017	1	3	3	3	2	3	1	5	3	3	2	3
ACE 018	2	3	2	2	2	3	1	5	3	1	1	3
ACE 020	3	3	3	3	1	1	1	5	3	1	2	2
ACE 021	1	3	3	3	2	3	3	4	3	1	2	2
ACE 022	1	3	2	3	2	3	3	5	3	3	2	3
ACE 024	2	3	2	2	1	3	1	6	5	1	1	3
ACE 025	1	3	3	3	2	2	1	4	3	1	1	1
ACE 026	3	3	3	3	2	2	1	5	3	1	1	2
ACE 027	2	3	2	3	2	3	3	6	5	1	2	3
ACE 028	1	3	3	3	2	3	1	4	3	1	1	3
ACE 029	1	3	2	3	2	3	1	4	3	3	2	2
ACE 030	2	3	3	3	2	3	1	5	3	1	2	3
ACE 031	1	3	2	3	2	3	1	4	3	3	1	3
ACE 034	3	3	3	3	1	3	1	4	3	1	2	2
ACE 035	3	3	3	3	2	3	1	4	5	1	1	2
ACE 036	3	3	3	3	2	1	1	4	3	1	1	3
ACE 037	1	3	3	3	2	3	1	4	3	3	2	3
ACE 038	3	3	3	3	2	3	3	5	3	3	1	2
ACE 039	3	3	3	3	2	2	1	4	3	3	1	2
ACE 042	2	3	3	3	2	1	1	4	3	3	2	2

ACE 043	2	3	3	3	2	3	1	4	3	1	1	2
ACE 044	1	3	3	3	1	3	3	4	3	3	1	3
ACE 045	1	3	2	3	1	3	3	5	3	1	1	3
ACE 046	3	3	3	3	2	3	3	5	3	1	1	3
ACE 047	2	3	3	3	2	1	1	5	3	1	2	2
ACE 048	3	3	3	3	2	3	3	4	3	1	1	2

= leaf shape (1-rounded; 2-ovate; or 3-elliptical); FGBF = general shape of the edges of the mature leaf (1- straight edge; 2- slightly wavy edge; 3- wavy edge); FL = general shape of the leaf limb (1- open; 2- intermediate; 3- closed); CLC = colour of the flower corolla lobes (1- white; 2- light pink; 3- dark pink; 4- violet); CCFI = colour of the skin of the immature fruit (1- green; 2- yellowish green; 3- purplish green; 4- greenish purple; 5- purple); CCFM = ripe fruit skin colour (1- yellow; 2- orange; 3- pink; 4- red; 5- purplish red; 6- purple); CPFM = ripe fruit flesh colour (1- white; 2- green; 3- yellow; 4- pink; 5- red; 6- purple); TC = rind texture (1- smooth; 2- intermediate; 3- wrinkled); SSF = fruit surface groove (1- superficial; 2- intermediate; 3- deep); TFM = mature fruit size (1- small fruit (12-16 mm); 2- medium fruit (17-23mm); 3- large fruit (24-30mm)).

The crown conformation descriptor showed 16 accessions with an upright shape, i.e. 42.00% of the total number of individuals had vertical crown growth.

For the descriptor crown branching, with the exception of ACE 002 and ACE 004, which are branched, and ACE 011, which is slightly branched, all the other accessions had a very branched crown. In percentage terms, around 92.66 per cent of the Pesagro-Rio acerola accessions had this type of branching.

In terms of leaf shape, we didn't find any individuals with rounded leaves, but most of the accessions had elliptical leaves (73.68%) or ovate leaves (26.32%).

With regard to the general shape of the leaf edges, there was a predominance of wavy edges (33 individuals), totalling 86.84% of the population. There was only one individual with a straight edge (ACE 004).

With regard to the general shape of the leaf limb, no accessions were found with a closed shape. More than half of the accessions had an intermediate shape (71.05%), while 28.95% of the accessions had an open shape.

The colouring of the flower corolla lobes varied from white to dark pink, and no violet colouring was identified among the accessions.

There was no significant variation in the colour of the skin of the immature fruit, as only two colours were found among all the accessions studied. Most of the accessions (72.30%) had purplish green skin colour and the rest (23.70%) were green.

The greatest diversity among the accessions was seen in the colour of the skin of the ripe fruit. The predominant colour was red, with 57.89% of the individuals, followed by purplish red, with 31.57% of the individuals, and purple, with 10.58% of the individuals.

The colour of the flesh of the ripe fruit was predominantly yellow in most of the accessions (78.95%). Seven individuals were red (18.45%) and only one was pink (2.6%).

The texture of the skin of most of the fruit was smooth (65.78%) and the rest was wrinkled (37.22%). No genotype with an intermediate texture was observed among the accessions.

The descriptor groove of the ripe fruit showed all three shapes, with 21 individuals having a superficial shape (55.26%), 15 individuals having an intermediate shape (39.47%) and only two individuals having a deep shape (5.26%).

For the descriptor mature fruit size, it was possible to observe the three types of size. More than half of the accessions had medium-sized fruit (55.26%). ACE 002 was the only one with a small size and the rest of the accessions had large fruit (41.80%).

4.1.2 Morpho-agronomic characterisation - quantitative characters

The results of the evaluation of the quantitative characters applied to the acerola accessions are shown in Table 2. Wide variability was detected among the accessions.

Table 2 - Averages of the 38 accessions for nine quantitative descriptors used for acerola.

Access	H (m)	SD (m)	DC (cm)	CF (cm)	LF (cm)	NO. MFP	N°FP	D (mm)	C (mm)
ACE 001	3,50	4,00	49,00	56,75	27,50	4,00	4,00	22,87	20,47
ACE 002	3,60	2,20	42,00	50,50	19,50	4,00	4,00	19,42	16,67
ACE 004	2,00	2,52	21,00	53,50	26,65	4,00	4,00	26,22	23,06
ACE 005	2,80	3,20	44,00	52,50	28,00	7,00	4,00	26,07	23,20
ACE 007	4,00	3,30	45,00	42,00	38,50	6,00	4,00	21,03	20,00
ACE 008	3,50	4,90	48,00	45,00	34,00	3,00	3,00	27,24	24,54
ACE 009	4,10	3,40	48,00	53,00	34,00	5,00	4,00	24,17	22,25
ACE 010	4,20	2,20	36,00	62,00	48,00	4,00	3,00	22,11	19,82
ACE 011	3,90	3,40	39,00	52,00	31,00	4,00	4,00	22,07	16,67
ACE 012	3,40	3,30	41,00	54,00	34,00	5,00	4,00	26,73	22,40
ACE 014	3,30	2,60	32,00	61,00	31,00	4,00	4,00	22,60	20,47
ACE 016	2,40	2,10	20,00	46,00	24,00	4,00	2,00	24,13	20,71
ACE 017	2,10	2,35	19,00	45,00	23,00	6,00	4,00	26,71	21,93
ACE 018	3,40	4,00	38,00	70,00	45,00	5,00	4,00	25,47	21,93
ACE 020	2,30	1,80	30,00	38,00	27,00	4,00	4,00	22,42	20,08
ACE 021	1,70	3,10	18,00	46,00	22,00	4,00	2,00	26,82	22,13
ACE 022	2,40	3,10	18,00	56,00	37,00	6,00	3,00	27,93	23,79
ACE 024	3,70	3,30	39,00	49,00	30,00	6,00	6,00	24,14	20,25
ACE 025	1,80	2,10	26,00	35,00	21,00	3,00	2,00	23,12	19,18
ACE 026	4,40	3,00	33,00	53,00	23,00	3,00	3,00	21,17	18,53
ACE 027	3,90	3,40	48,00	68,00	34,00	6,00	4,00	25,30	21,43
ACE 028	3,30	3,80	40,00	48,00	25,00	6,00	4,00	25,75	21,75
ACE 029	3,00	4,80	31,00	50,00	32,00	6,00	3,00	24,47	21,69
ACE 030	3,50	4,90	49,00	43,00	30,00	6,00	4,00	23,43	20,07
ACE 031	2,30	4,00	25,00	41,00	27,00	6,00	4,00	23,80	20,10
ACE 034	3,70	4,10	55,00	56,00	40,00	6,00	5,00	23,73	20,20
ACE 035	3,90	2,80	40,00	58,00	34,00	5,00	4,00	22,17	19,00
ACE 036	1,90	1,50	17,00	43,00	20,00	3,00	3,00	23,42	21,25
ACE 037	3,10	3,10	40,00	50,00	19,00	6,00	4,00	23,75	19,00
ACE 038	1,80	1,90	14,00	36,00	23,00	5,00	4,00	21,71	18,86
ACE 039	2,70	2,80	37,00	46,00	27,00	6,00	3,00	21,70	18,07
ACE 042	2,20	2,20	50,00	47,00	20,00	3,00	2,00	24,67	19,00
ACE 043	2,30	2,70	20,00	52,00	24,00	6,00	4,00	24,97	20,35
ACE 044	3,20	3,10	55,00	47,00	24,00	6,00	2,00	26,25	21,89
ACE 045	3,60	3,20	58,00	47,00	31,00	6,00	3,00	25,97	20,90
ACE 046	3,40	4,60	45,00	37,00	24,00	7,00	3,00	26,07	22,50
ACE 047	2,60	2,70	37,00	42,00	25,00	3,00	3,00	22,18	19,29
ACE 048	2,30	1,70	22,00	41,00	20,00	4,00	3,00	22,80	19,90

H(m)= plant height; DP (m)= plant diameter; DC (cm) = stem diameter; CF (cm) = mature leaf length; LF (cm) = maximum leaf width; N°MFP = average number of flowers per panicle; N°FP = average number of fruits per panicle; D (mm) = fruit diameter; C (mm) = fruit length.

Plant height ranged from 1.70 metres to 4.40 metres. Accession ACE 026 had the highest plant height (4.40 m), followed by accessions ACE 010 (4.20 m) and ACE 009 (4.10 m). The lowest heights were found among accessions ACE 021 (1.70 m), ACE 025 and ACE 039, both with 1.80 m, and ACE 035 (1.90 m).

For the stem diameter characteristic, two accessions showed the same length, standing out as the largest: access ACE 008 and ACE 030, both with 4.90 metres. The second longest was ACE 029, with a length of 4.80 metres. In addition to its low height (1.90 m), access ACE 36 had the smallest stem diameter of all, at 1.50 metres.

Three accessions stood out with the same stem diameter, ACE 008, ACE 009 and ACE 027, all with 48.00 centimetres. However, it was accessions ACE 001 and ACE 030 that had the largest diameters, at 49.00 centimetres.

Accession ACE 018 stood out with the greatest leaf length (70 cm) and the second greatest maximum leaf width (45.00 cm). Accession ACE 010 was the highlight for the maximum leaf width variable, with 48.00 cm and the third greatest leaf length, with 62.00 centimetres. The lowest values were found in accession ACE 037 (19.00 cm) and accession ACE 002 (19.50 cm).

Two accessions had an average of seven leaves per panicle, culminating in four fruits at the end of fruiting. These accessions were ACE 005 and ACE 046. Most of the accessions had between three and six leaves per panicle. The accessions ACE 016, ACE 021, ACE 042 and ACE 044 had the lowest number of fruits per panicle with only two fruits each.

Several accessions had fruit diameters in the 26.00 millimetre range (ACE 004, ACE 005, ACE 012, ACE 017, ACE 021, ACE 044 and ACE 046). In addition, the largest diameter was observed in accession ACE 022, with a value of 27.92 millimetres. The smallest fruit diameter and length were found for accession ACE 002, with 19.42 and 16.67 millimetres, respectively. These results are satisfactory, given that, according to the Brazilian Fruit Institute (IBRAF,1995), among the quality criteria established for acerola, the fruit must have a minimum diameter of 15 mm in order to be accepted by the processing industries. All the accessions studied exceeded this standard.

4.2 Main Components

The results obtained for the principal components, the eigenvalues and the percentages of the variance explained by the components obtained from nine quantitative characteristics are shown in

Table 3.

Table 3 - Estimates of variance (eigenvectors, root), cumulative percentage of nine quantitative descriptors, evaluated in 38 acerola accessions from Pesagro-Rio.

Principal component (PC)	Root	Root (%)	Accumulated %
CP1	3,26	36,24	36,24
CP2	2,15	23,86	60,10
CP3	1,12	12,52	72,63
CP4	1,02	11,39	84,02
CP5	0,45	5,03	89,06
CP6	0,41	4,56	93,62
CP7	0,30	3,43	97,05
CP8	0,17	1,98	99,04
CP9	0,86	0,95	100,00

In genetic divergence studies between a group of genotypes, it is desirable for the accumulated variance in the first two principal components to exceed 80% (Cruz & Regazzi, 1994). However, from the results obtained, the accumulated variance between the components did not show a concentrated distribution in the first two components. In this study, it was only possible to accumulate a satisfactory percentage from the first four components, with 84.07% of the total variability of the accessions. The first component explained 36.24%, the second 23.86%, the third 12.52% and the fourth 11.39% of the total variation. The other components together accounted for only 15.97%.

Martinello et al. (2003), studying genetic divergence in okra accessions based on 13 quantitative morphological characters, found similar results to those observed in this study. The authors concluded that four components were necessary for the variance explained by them to reach a minimum of 80%. For the authors, these data are indicative of a more equitable dispersion of the total variance in the characters evaluated.

Rosa et al. (2006), carrying out agromorphological characterisation on Oryza Glumaepatula with the aid of principal component analysis using 15 characters, observed that only the first two components (57.8%) were not enough to explain the total variance. It was necessary to use the variance accumulated by the first four components to reach 84.9% of the total variation observed. There was great variability and differentiation for morphological and agronomic characters between populations of O. glumaepatula from different Brazilian river basins.

Principal component analysis makes it possible to discard variables, thus eliminating those that make little contribution to the divergence study. To discard variables, we adopted Jolliffe's (1973) recommendation that the number of variables discarded should be equal to the number of components whose variance (eigenvalue) was less than 0.7. Of the nine principal components, five (55.5% of the characters) had a variance lower than the established eigenvalue. As a result, the variables with the highest coefficients, starting with the last principal component, were discarded. These were height, plant diameter, average number of leaves per panicle, number of fruits per panicle and fruit diameter.

4.3 Relative importance of characters

Using Singh's method (1981), it was determined that of the nine quantitative traits studied, three contributed 96.03% to genetic divergence, while six traits contributed only 3.97%. The stem diameter character showed the greatest relative contribution (54.12%) in relation to the quantitative characters, followed by mature leaf length (24.00%) and maximum leaf width (17.19%), as can be seen in Table 4.

Table 4 - Relative importance of nine quantitative characteristics evaluated in 38 acerola accessions obtained using Singh's method (1981).

Variable	S.j	Value in %
H	849,76	0,218
DP	1127,78	0,289
DC	211013,0	54,128
CF	93595,81	24,009
LF	69841,03	17,915
NO. MFP	2157,0	0,553
N°	1044,0	0,267
D	5723,97	1,468
C	4483,90	1,150

H = plant height; DP = plant diameter; DC = stem diameter; CF = mature leaf length; LF = maximum leaf width; N°MFP = average number of flowers per panicle; N°FP = average number of fruits per panicle; D = fruit diameter; C = fruit length.

Of the nine descriptors studied, only three (33.33%) were selected as the most important for studying genetic diversity between acerola accessions.

Blank et al. (2004), studying the morphological and agronomic characterisation of accessions of basil and lavender, found that there was great diversity between the accessions, using the descriptors stem diameter, leaf length and leaf width, showing that they can always be used in characterisation studies.

Daher et al. (1997), in a study of genetic divergence between elephant grass accessions, used the principal components technique and observed that, out of a total of 22 characters evaluated over three years, only eight (36.4%) were selected as the most important for determining genetic divergence.

4.4 Grouping methods

4.4.1 Tocher Optimisation Method

Cluster analysis using the Tocher technique (Table 5), based on the dissimilarity matrix based on the average Euclidean distance, revealed the emergence of 13 groups.

Table 9 - Groups of accessions established by the Tocher method, based on genetic dissimilarity between nine quantitative descriptors for 38 chard accessions.

Groups	Access
I	ACE 036 (28), ACE 048 (38), ACE 016 (12), ACE 025 (19), ACE 020 (15), ACE 047 (37), ACE 038 (30)
II	ACE 017 (13), ACE 043 (33), ACE 004 (03), ACE 031 (25), ACE 028 (22), ACE 005 (04), ACE 012 (10), ACE 037 (29), ACE 029 (23)
III	ACE 014 (11), ACE 035 (27), ACE 001 (01), ACE 009 (07), ACE 027 (21), ACE 034 (26), ACE 011 (09), ACE 007 (05), ACE 030 (24)
IV	ACE 044 (34), ACE 045 (35), ACE 046 (36)
V	ACE 002 (02), ACE 023 (20)
VI	ACE 039 (31)
VII	ACE 042 (32)
VIII	ACE 010 (08)
IX	ACE 024 (18)
X	ACE 021 (16)
XI	ACE 018 (14)
XII	ACE 008 (06)
XIII	ACE 022 (17)

The first group consisted of seven individuals. This group included the accession with the smallest plant diameter, 1.50 metres (ACE 036). In addition to this, the accessions ACE 048 (38), ACE 016 (12), ACE 025 (19), ACE 047 (37), ACE 020 (15) and ACE 038 (30) also appeared. The second and third groups were the most numerous, made up of nine materials each. Group II included accessions 13, 33, 25, 22, 04, 10, 29 and 23. Group III consisted of accessions 11, 27, 01, 07, 21, 26, 09, 05 and 24. Group IV consisted of three accessions, 34, 35 and 36, while group V had two accessions, 02 and 20.

For the other groups (VI, VII, VIII, IX, X, XI, XII and XIII), all had just one accession each, 31, 32, 08, 18, 16, 14, 06 and 17, respectively. The main highlight was group XIII, with access 17 (ACE 022), which had the largest fruit diameter of all the other accessions.

The quantitative characteristic that most contributed to the formation of groups by Tocher analysis was stem diameter. This can be affirmed since accessions with similar sizes of this descriptor formed groups, and between groups these sizes were different.

4.4.2 UPGMA hierarchical method

Based on the data generated by the dissimilarity matrix, the dendrogram with the clusters shown in Figure 1 was obtained using the UPGMA clustering method, where the Y axis shows the percentage distances between the accessions and the X axis shows the 38 accessions, based on nine quantitative characteristics. Hierarchical clustering methods allow groups to be established in such a way that there is homogeneity within the group and heterogeneity between groups. To decide on the number of groups formed in the dendrogram, a cut-off point was adopted at the location of the highest level change (90%). This cut resulted in the formation of 13 groups: group I (accessions 20, 09, 02, 27, 11 and 01); group II (accessions 36, 35, 34, 10, 07, 22 and 04); group III (accessions 25 and 23); group IV (accessions 31 and 29); group V (accessions 26 and 18); group VI (accessions 24 and 05); group VII (accesses 08); group VIII (accesses 21 and 14); group IX (accesses 33, 13 and 03); group X (access 17); group XI (access 16); group XII (accesses 37, 32, 30, 15, 19, 18, 38, 28 and 12) and group XIII (access 06).

Comparing the Tocher optimisation method with the UPGMA hierarchical grouping, partial agreement was found in the results. Both the Tocher and UPGMA methods found that accessions 37, 38, 30, 28, 19, 15 and 12 were in the same group. Accessions 6, 8, 16 and 17 were isolated in both groupings. Sudré et al. (2006), studying genetic divergence between chilli pepper accessions, found similar results, as there was partial agreement between the two grouping methods.

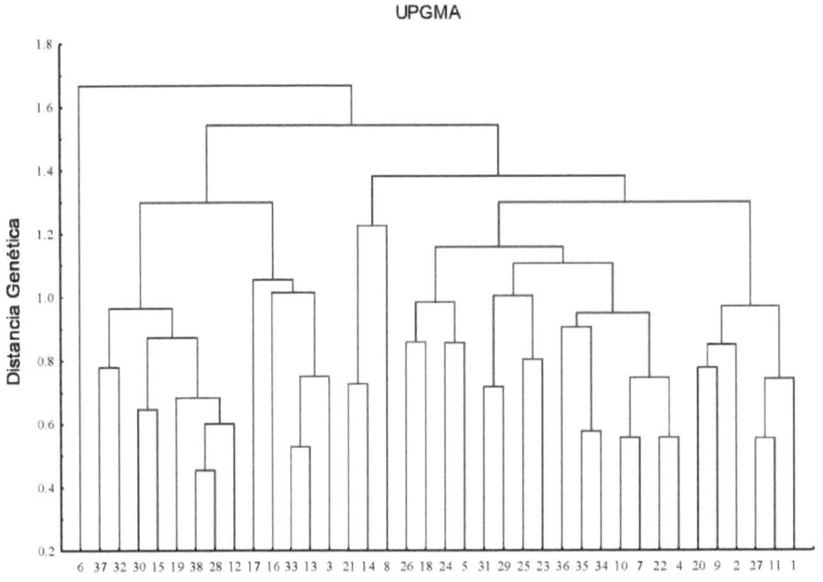

Figure 1 - Dendrogram established from the dissimilarity pattern quantified by UPGMA, of 38 acerola accessions.

4.4.3 Multicategorical variables

The Tocher optimisation method, based on the average Euclidean distance, revealed the formation of 10 groups. Group I comprised 42.10 per cent of the accessions and the other accessions were distributed in the remaining groups (Table 6). In this group, all the accessions had a highly branched canopy, wavy leaf edges and an intermediate leaf limb shape.

Table 6 - Groups of accessions established by the Tocher method, based on genetic dissimilarity between the 17 qualitative descriptors for acerola.

Groups	Access
I	ACE 012 (10), ACE 014 (11), ACE 039 (31), ACE 029 (23), ACE 037 (29), ACE 017 (13), ACE 031 (25), ACE 028 (22), ACE 043 (33), ACE 021 (16), ACE 034 (26), ACE 048 (38), ACE 035 (27), ACE 036 (28), ACE 025 (19)
II	ACE 016 (12), ACE 024 (18), ACE 001 (01)
III	ACE 020 (15), ACE 047 (37), ACE 023 (20), ACE 030 (24), ACE 046 (36), ACE 038 (30)
IV	ACE 005 (04), ACE 008 (06), ACE 018 (14)
V	ACE 022 (17), ACE 045 (35), ACE 044 (34)
VI	ACE 007 (05), ACE 010 (08)
VII	ACE 009 (07), ACE 027 (21)
VIII	ACE 011 (09)
IX	ACE 002 (02)
X	ACE 004 (03)

Groups II (accessions ACE 016, ACE 024, ACE 001), IV (accessions ACE 005, ACE 008, ACE 018) and V (accessions ACE 022, ACE 045, ACE 044) added three accessions each.

All the accessions in group II had an open leaf limb shape, a green colour on the skin of the immature fruit, a purple colour on the skin of the ripe fruit, a whitish colour on the flesh of the ripe fruit and a shallow groove on the surface of the fruit.

For group IV, the accessions were homogeneous in terms of the general shape of the leaf limb, all intermediate, the colour of the corolla lobes of the flowers was light pink, the colour of the skin of the immature fruit was green, the colour of the skin of the ripe fruit was purplish red, the texture of the skin was smooth and the fruit was large.

The accessions in group V had a globular crown shape, a highly branched crown, the general shape of the wavy edges of mature leaf, the colour of the corolla lobes of the flowers was dark pink, the colour of the skin of the immature fruit was purplish green and the fruit was large.

The genetic dissimilarity measures made it possible to estimate that the most differentiated accessions were: ACE 002 (2) and ACE 024 (18); ACE 007(5) and ACE 021 (16); ACE 011 (9) and ACE 022 (17), as they showed the maximum dissimilarity value, being 0.647 %, 0.529 % and 0.538 %, respectively.

The smallest magnitudes of distance were between the accessions: ACE 012 (10) and ACE 014 (11); ACE 012 (10) and ACE 021(16); ACE 017(13) and ACE 027 (21), with 0.058% dissimilarity, identical for the pairs of accessions grouped together.

The formation of the groups did not follow just one specific characteristic that could determine the separation between the accessions, but eight descriptors were crucial in determining the most distant accessions. ACE 002 had CC (erect), FF (elliptical), FL (closed), CCFM (light pink), CPFM (purplish red), TC (wrinkled) and SSF (deep). On the other hand, accession ACE 024 showed CC (intermediate), FF (ovate), FL (open), CCFM (dark pink), CPFM (purple), TC (smooth) and SSF (shallow).

The dendrogram with the clusters shown in Figure 2 was obtained using the UPGMA clustering method, based on the data generated by the dissimilarity matrix between the 38 accessions.

To decide on the number of groups formed in the dendrogram, a cut-off point was adopted at the location of the highest level change (25%). This cut resulted in the formation of 10 groups: group I (accessions 01, 05); group II (accessions 09, 26, 15 and 08); group III (accessions 32, 31, 11, 10, 20 and 02); group IV (accessions 36, 30, 28, 22, 29, 33, 27, 38 and 16); group V (accessions 37, 24 and 07); group VI (accessions 25, 23, 17, 29 and 13); group VII (accessions 35 and 34); group VIII (accessions 14, 06 and 04); group IX (accessions 21, 18 and 12) and group X (accession 03). There was partial agreement between the grouping methods.

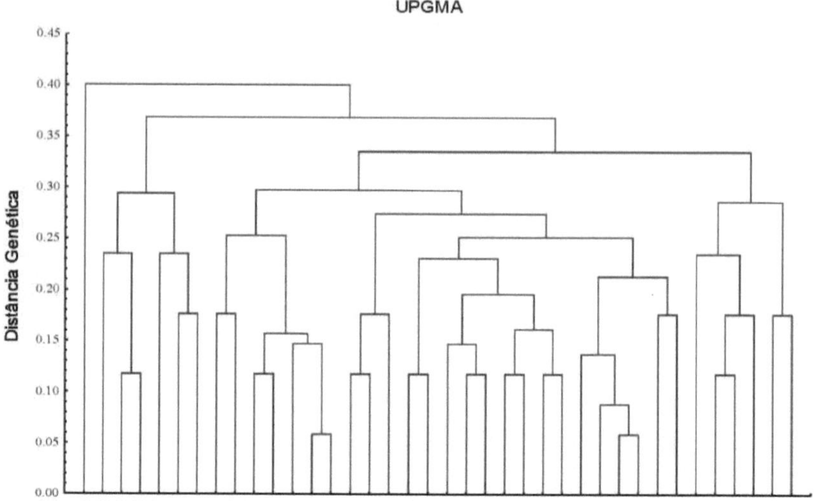

Figure 2 - Dendrogram established from the dissimilarity pattern quantified by UPGMA, of 38 acerola accessions.

The Tocher and UPGMA optimisation methods were in agreement and satisfactory for forming fairly homogeneous groups for the characteristics assessed. According to Vieira et al. (2005), establishing groups with genotypes that are homogeneous within and heterogeneous between groups is the starting point for a more detailed evaluation of these genotypes in order to use them in breeding programmes.

Bento et al. (2007) studied qualitative and multicategorical descriptors in the estimation of phenotypic variability between chilli pepper accessions and concluded that the multicategorical analysis carried out with 28 accessions allowed seven groups to be formed using the Tocher method. According to the authors, the analysis of multicategorical variables proved to be efficient in grouping the pepper accessions studied, indicating that its use in quantifying phenotypic divergence and identifying heterotic groups can help in the management of the germplasm bank and in the selection of accessions for genetic improvement programmes.

Sudré et al. (2006), when studying phenotypic divergence between 59 Capsicum accessions, using 15 qualitative descriptors and multicategorical data, obtained the formation of eight groups using the Tocher method. According to the authors, the formation of these clusters between accessions with

This would prove the effectiveness of this method for grouping accessions with a small genetic distance between them. The authors suggest that collecting multicategorical data is more economical and less time-consuming than molecular data.

4.5 Physical, chemical and physical-chemical variables

Table 7 shows the average values for fruit diameter, length and shape index.

Table 7 - Averages of six physical characteristics for the 25 acerola accessions from Pesagro. Ripening stages: 1= early; 2= semi-ripe; 3= ripe. The data represents the average of 20 fruits from each access.

	Diameter (mm)			Length (mm)			Format index (mm)		
Access	1	2	3	1	2	3	1	2	3
ACE 001	20,19	21,50	22,87	18,63	19,25	20,47	1,08	1,12	1,12
ACE 002	17,07	16,76	19,42	14,86	15,00	16,67	1,15	1,12	1,17
ACE 008	25,29	27,14	27,24	23,25	24,93	24,54	1,09	1,09	1,11
ACE 009	22,75	23,33	24,17	21,42	21,47	22,25	1,06	1,09	1,09
ACE 010	20,92	20,08	22,11	18,75	18,67	19,82	1,12	1,08	1,12
ACE 011	23,17	24,44	22,07	19,75	20,08	16,67	1,17	1,22	1,32
ACE 012	23,25	26,08	26,73	21,58	22,17	22,40	1,08	1,18	1,19
ACE 014	21,08	21,43	22,60	19,08	19,71	20,47	1,10	1,09	1,10
ACE 018	23,50	24,85	25,47	20,75	20,87	21,93	1,13	1,19	1,16
ACE 020	21,75	22,07	22,42	19,67	20,50	20,08	1,11	1,08	1,12
ACE 022	24,33	26,92	27,93	21,83	23,67	23,79	1,11	1,14	1,17
ACE 024	21,12	23,03	24,14	18,81	20,50	20,25	1,12	1,12	1,19
ACE 026	18,92	19,67	21,17	16,67	17,17	18,53	1,14	1,15	1,14
ACE 027	22,29	23,04	25,30	19,25	19,50	21,43	1,16	1,18	1,18
ACE 029	21,25	21,32	24,47	20,00	19,86	21,69	1,06	1,07	1,13
ACE 030	18,77	21,57	23,43	16,86	18,89	20,07	1,11	1,14	1,17
ACE 031	21,46	22,35	23,80	17,88	18,75	20,10	1,20	1,19	1,18
ACE 036	21,33	22,64	23,42	20,17	20,43	21,25	1,06	1,11	1,10
ACE 037	21,58	23,50	23,75	18,75	19,83	19,00	1,15	1,18	1,24
ACE 038	20,32	20,50	21,71	17,35	17,64	18,86	1,17	1,16	1,15
ACE 042	21,25	25,50	24,67	17,50	20,70	19,00	1,21	1,23	1,30
ACE 043	24,33	24,89	24,97	20,71	20,43	20,35	1,18	1,22	1,23
ACE 044	24,00	25,88	26,25	21,17	21,67	21,89	1,13	1,19	1,20
ACE 045	20,70	23,79	25,97	17,53	19,82	20,90	1,18	1,20	1,24
ACE 046	23,35	24,82	26,07	20,40	21,68	22,50	1,14	1,14	1,16

The physical evaluations showed that as the ripening process progressed there was an increase in the diameter and length values. These results are in line with the literature (França and Narain, 2003; Chitarra, 1994; Nogueira, 1997). Fruit diameter ranged from 17.07 mm to 27.93 mm and length from 14.86 mm to 24.55 mm, with the highest values being observed in the ACE 008 and ACE 022 accessions.

The shape index, which reflects the relationship between the diameter and length of the fruit, ranged from 1.06 to 1.24. According to Brunini et al. (2004), acerola fruit with a shape index between 0.86 and 1.24 confirms that this crop is a drupe subglobulose.

Fruit skin colour measurements (Table 8) were taken for fruit at three stages of ripeness using a colorimeter (Chroma Meter, model CR-300, Minolta). For this purpose, three readings were taken in equidistant regions, which made up an average value for the following colour parameters: i) luminosity: Hunter's parameter (L); ii) chromaticity: Hunter's parameter (a), which indicates the change from green to red and Hunter's parameter (b), which indicates the evolution from blue to yellow; hue angle (h), which indicates the colour of the sample.

Table 8 - Values of the colourimetric components L*, a* and b* h° of fruit from 25 Pesagro-RIO accessions at three stages of ripeness.

	L*			a*			b*			h°		
Access	1	2	3	1	2	3	1	2	3	1	2	3
ACE 001	57.58	48.87	33.63	13.80	47.14	36.67	35.82	33.80	13.78	69.10	35.50	20.50
ACE 002	53.33	32.38	38.34	20.71	39.84	46.56	34.70	34.24	16.01	60.65	23.10	18.90
ACE 008	48.50	39.30	30.30	16.40	39.33	35.50	31.50	23.13	13.50	62.68	31.90	20.20
ACE 009	64.50	57.40	44.30	16.29	34.35	50.38	45.46	40.16	19.00	70.20	49.90	20.04
ACE 010	60.49	48.37	42.74	20.34	35.57	48.84	44.62	33.91	24.82	65.45	44.00	26.85
ACE 011	61.03	42.04	43.16	9.96	41.49	47.79	44.10	24.93	17.12	77.15	31.00	19.60
ACE 012	53.45	39.64	38.89	14.27	38.60	40.46	28.55	22.44	14.51	63.50	30.20	16.30
ACE 014	63.42	57.02	37.31	9.15	23.44	40.54	41.86	38.13	19.38	77.75	58.60	25.35
ACE 018	53.80	38.65	30.48	26.76	45.17	26.52	33.72	20.57	24.04	51.60	24.45	14.97
ACE 020	79.98	50.02	44.76	(0.34)	36.36	47.47	52.83	29.31	18.35	90.44	39.00	21.00
ACE 022	39.93	37.15	29.44	14.71	35.72	30.89	19.56	24.26	9.94	53.10	33.65	17.70
ACE 024	45.82	42.04	33.85	16.00	38.39	39.86	24.52	26.02	16.40	56.85	34.10	21.54
ACE 026	65.24	47.06	37.07	9.46	45.52	42.94	41.70	32.65	20.58	77.30	33.50	25.50
ACE 027	55.36	51.72	36.14	19.59	37.79	43.40	31.93	32.60	19.24	58.45	40.78	23.75
ACE 029	44.10	32.33	23.36	18.88	39.14	25.13	24.09	13.07	5.46	51.30	20.80	12.20
ACE 030	52.04	40.38	31.47	23.20	40.08	34.96	32.13	22.46	10.37	54.25	28.95	16.45
ACE 031	67.69	61.17	38.94	(2.05)	19.65	42.60	50.66	47.63	22.84	92.10	67.85	28.15
ACE 036	77.80	57.22	41.40	9.30	51.70	44.00	48.60	31.40	13.88	79.20	31.50	17.44
ACE 037	77.00	47.30	37.60	(6.25)	50.26	39.20	39.50	23.76	12.23	85.30	26.30	16.70
ACE 038	78.63	54.29	37.70	(5.30)	50.80	39.31	38.70	29.52	10.25	93.30	30.50	14.50
ACE 042	78.50	45.50	41.22	(6.60)	39.50	43.50	42.50	23.35	15.25	89.10	31.20	19.60
ACE 043	53.97	52.61	39.37	17.13	33.63	43.71	29.41	33.66	27.73	71.50	44.75	27.00
ACE 044	70.99	49.82	33.38	8.28	33.67	37.28	34.43	32.82	12.49	76.40	43.80	18.40
ACE 045	51.22	48.46	33.57	22.22	40.14	39.49	32.20	34.87	15.96	56.30	40.85	21.90
ACE 046	51.30	44.40	31.67	18.30	38.00	32.51	31.50	30.50	8.90	60.70	36.50	15.30

Ripening stages: 1= once; 2= semi-ripe; 3= ripe. The data represents the average of 20 fruits from each access

For all the accessions, the luminosity (L*) of the fruit decreased from the **'once' stage to the semi-ripe stage and remained decreasing until the** fruit was fully **ripe**. These results are in line with previous observations indicating that the L* value decreases with the appearance of red colour as the fruit ripens, representing the loss of fruit brightness due to the synthesis of carotenoids and a decrease in green colour (Lopez Camelo; Gomez, 2004). The L* values were higher among the ACE 020, ACE 038 and ACE 042 accessions and lower among the ACE 030, ACE 022 and ACE 029 accessions, respectively.

The values of the chromatic component a* (+a*: degree of red colour of the fruit; -a*: degree of green colour), in 52% of the accessions, increased as the **fruit went from the 'once' stage (some with negative values) to** semi-ripe and ripe (positive values) as a result of anthocyanin synthesis and chlorophyll degradation (Conceição, 1997). On the other hand, 48% of the **accessions showed an increase from the 'once' stage to the semi-mature stage and** a decrease to the mature stage.

The values of the chromatic component b* (+b*: degree of yellow colour; -b*: degree of blue colour) showed a tendency to decrease until the semi-ripe and ripe stages, corresponding to a marked synthesis of anthocyanins (red colour) **and other compounds such as** p-carotene (orange colour), which mask the yellow pigments (Lima et al., 2003; Lima et al., 2005). Carvalho et al. (2005) estimated the lycopene content in the fruit of tomato genotypes by colourimetric analysis and concluded that the sharp decrease in the b* component, causing the change to red colour in tomatoes, is due to the synthesis of lycopene, which masks the yellow colour.

The H* data showed a decrease from the initial stage to the end of ripening. Lima et al. (2007), studying the correlation between anthocyanin content and the chromatic characterisation of pulp from different acerola genotypes, stated that the lower the angle*, the closer to the a* axis, the redder the pulp and skin of the fruit. According to the authors, this is due to the increased content of anthocyanin pigments.

Accession ACE 029, at thc mature stage, had a luminosity value of L* (23.36), a* parameter values (25.13), b* parameter value (5.42) and H* (12.20), characterising a dark, purplish-red, intense colour. Although 60% of the accessions showed a dark colour at the last stage of ripeness, accessions ACE 038, ACE 018 and ACE 046 were considered the darkest because they had the lowest H* values, 14.50, 14.97 and 15.30, respectively.

Table 9 shows the variation in four chemical characteristics (pH, ATT, SS and ascorbic acid content) obtained from acerola fruit at three stages of ripeness.

Table 9 - Averages of four chemical characteristics for the 25 acerola accessions from Pesagro.

	PH			ATT			SST			Vit C		
Access	1	2	3	1	2	3	1	2	3	1	2	3
ACE 001	3,52	3,40	3,34	0,80	0,72	0,48	7,10	8,10	9,90	3177,54	2723,82	2158,52
ACE 002	3,87	3,89	3,91	0,35	0,35	0,19	8,00	9,50	9,50	1868,44	1533,72	1444,46
ACE 008	3,09	3,15	3,27	1,02	1,17	0,95	7,20	7,80	7,80	2307,28	2277,53	1161,82
ACE 009	3,48	3,50	3,40	0,69	0,69	0,60	8,00	8,00	8,10	2232,90	2024,64	1935,38
ACE 010	3,28	3,34	3,31	0,65	0,58	0,36	6,50	7,00	8,90	1481,66	1384,96	1132,06
ACE 011	3,74	3,78	3,64	0,51	0,48	0,28	8,20	8,80	10,60	2069,27	1727,11	1273,39
ACE 012	3,48	3,49	3,39	0,71	0,65	0,36	7,00	8,50	9,40	2277,53	1608,10	1667,61
ACE 014	3,51	3,50	3,44	0,79	0,79	0,71	7,00	7,00	7,00	2798,20	2500,67	2218,03
ACE 018	3,35	3,37	3,40	0,92	0,88	0,32	6,90	6,90	8,50	3177,54	2389,10	1429,59
ACE 020	3,48	3,52	3,56	0,65	0,49	0,27	7,20	7,30	8,10	2113,89	1920,50	1890,75
ACE 022	3,33	3,32	3,32	0,94	0,93	0,44	7,10	8,20	8,80	2708,94	2619,68	1667,61
ACE 024	3,37	3,35	3,38	0,73	0,64	0,59	7,10	8,00	8,00	2203,15	1503,97	1459,34
ACE 026	3,37	3,29	3,38	1,15	0,89	0,77	6,90	6,90	8,00	2463,48	1801,49	1727,11
ACE 027	3,27	3,38	3,34	0,98	0,71	0,73	8,10	8,60	9,60	2396,54	2024,64	1280,83
ACE 029	3,25	3,25	3,25	0,94	0,96	0,96	7,00	7,40	9,10	2351,91	2039,51	1280,83
ACE 030	3,56	3,30	3,54	0,60	0,66	0,63	7,00	7,20	8,20	1920,50	1593,23	1429,59
ACE 031	3,10	3,08	2,63	1,14	1,18	1,09	6,00	6,90	8,75	2991,59	2738,69	2575,06
ACE 036	3,48	3,41	3,57	0,61	0,59	0,55	7,95	8,10	9,50	1905,63	1637,86	1637,86
ACE 037	3,33	3,29	3,36	0,75	0,69	0,65	8,00	8,00	8,20	2560,18	1875,87	1459,34
ACE 038	3,34	3,31	3,27	0,75	0,75	0,82	8,50	10,20	10,20	2946,96	2247,78	2024,64
ACE 042	3,55	3,51	3,51	0,52	0,35	0,35	7,80	9,20	11,00	2396,54	2374,23	1980,01
ACE 043	3,35	3,38	3,44	0,83	0,85	0,63	7,50	8,95	10,10	2649,44	2128,77	2024,64
ACE 044	3,21	3,29	3,28	1,04	0,88	0,77	7,00	7,20	7,80	3601,51	2322,16	1771,74
ACE 045	3,13	3,10	3,10	0,94	1,02	1,02	7,50	7,60	8,10	2619,68	1935,38	1161,82

ACE 046	3,25	3,25	3,25	1,02	0,97	0,97	7,10	7,50	7,30	3467,63	2619,68	2470,92

Ripening stages: 1= once; 2= semi-ripe; 3= ripe. The data represents the average of 20 fruits from each access

The pH of the fruit ranged from 3.09 to 3.91, values similar to those found by França and Nairan, (2003); Batista et al. (2000); Musser et al. **(2004). There was an increase in pH between the accessions from the 'once' stage to the** semi-mature stage. This result is perfectly understandable, as this increase in pH represents the decrease in fruit acidity as the fruit matures. However, as the fruit moved into the ripe stage, there was a slight decrease in the pH value in some accessions (ACE 001, ACE 009, ACE 010, ACE 011, ACE 012, ACE 014, ACE 027 and ACE 038).

The titratable acidity of the fruit decreased throughout the ripening stages, as was expected. ATT values ranged from 0.19 to 1.18. Musser et al. (2004) found acerola fruit acidity expressed as malic acid in the winter/1999 harvest, detecting a range of 1.31 to 2.04g of malic acid/100g of pulp. Although acerola is considered an acidic fruit, as it is rich in organic acids, with acidity percentages ranging from 0.65 to 1.68 (Matsuura et al., 2003), in this population (accessions ACE 002, ACE 010, ACE 011 and ACE 012), fruit with acidity compared to passion fruit was observed, ranging from 0.49 to 0.39% acidity.

Chaves et al. (2004), in a study on the physico-chemical characterisation of acerola juice, state that °Brix is used in the agro-industry to intensify control of the quality of the final product, control of processes, ingredients and others, such as: sweets, juices, nectar, pulp, condensed milk, alcohol, sugar, liqueurs and drinks in general, ice cream, among others. According to the authors, total soluble solids (°Brix) are used as a maturity index for some fruits and indicate the amount of substances that are dissolved in the juice, most of which are sugars.

In this study, the total soluble solids content increased as the fruit matured, ranging from 6.00 to 11.00 ° Brix. The fruit from accessions ACE 042, ACE 011 and ACE 038 had the highest percentages of TSS, with 11.00, 10.60 and 10.20, respectively, at the ripe stage. The lowest percentages were found in accessions ACE 031 and ACE 018, with 6.00 and 6.50 °Brix, respectively, at the 'once' stage. These results are in line with those observed by Alves (1996). According to the author, values of 5.00, up to a maximum of 12.00 °Brix can be found in acerola, with the average being around 7.00 to 8.00 °Brix. Gomes et al. (2000) found average values between 5.25 and 8.58 °Brix. Oliveira et al. (2007) found values between 6.70 and 9.40 °Brix.

Vitamin C contents ranged from 3467 mg ('once' stage) to 1161 mg/100g of pulp ('ripe' stage), with accessions ACE 046 and ACE 008, respectively. Four accessions had values higher than 3000 **mg/100g of pulp at the 'once' stage: ACE 001, ACE 014,**

ACE 046 and ACE 048, with 3177.54, 3177.54, 3601.51 and 3467.63 mg/100g of pulp, respectively. The lowest values for vitamin C content were observed for accessions ACE 008, ACE 045, ACE 010 and ACE 011, with 1161.82, 1168.82, 1132.06 and 1273.39 mg/100g of pulp, respectively, at the ripe stage.

The vitamin C content was reduced from the initial stage to the ripe stage with a loss of up to 45% due to biochemical oxidation. Vendramini and Trugo (2000), studying the chemical composition of acerola at three stages of ripeness, observed a 50% loss in vitamin C content throughout the ripeness stage. According to the authors, this decrease is due to the appearance of the chemical compound 3-hydroxy-2-pyrone during fruit ripening, which results in the oxidative breakdown of ascorbic acid. In addition, Butt (1980), cited by Nogueira et al., (2002) attributes this decrease in vitamin C content to the action of the enzyme ascorbic acid oxidase (ascorbate oxidase), which due to its higher activity in unripe fruit, would explain the losses due to ripening (Asenjo et al., 1960 cited by Nogueira et al., 2002).

Fruits in the 'once' stage showed the highest values of vitamin C content, regardless of the access, with a gradual reduction until the final ripeness of the fruit. These results indicate that there is a clear correlation between vitamin C content and stage of ripeness, in agreement with the literature (Santos et al., 1999; Carvalho, 1992; Sanches et al., 1996; Batista et al., 2000; Silva et al., 2007).

Nogueira et al. (2002) studied the effect of fruit ripeness on the physico-chemical characteristics of acerola and found that two of the accessions studied produced fruit with vitamin C levels suitable for both the domestic and foreign markets. Unripe fruit had significantly higher vitamin C levels than ripe and semi-ripe fruit, and could be used by the pharmaceutical industry.

Batista et al. (2000), in a study of the physical and chemical parameters of acerola (Malpighia punicifolia, l.), at different stages of ripeness, observed that the vitamin C content present in the material classified as green exceeded the content determined in the ripe samples by more than three times. The vitamin C content in the half-ripe material proved to be more than 1.8 times higher than the content in the ripe material and around 1.7 times lower than the content determined in the green material, demonstrating a clear correlation between the vitamin C content and the stage of ripeness of the acerola fruit.

4.6 . Main Components

The variance estimates and eigenvectors of each component obtained from the principal component analysis of 25 acerola accessions are shown in Table 10.

Table 10 - Estimates of the variances and eigenvectors of each component obtained from the principal component analysis of 25 acerola accessions from Pesagro-Rio.

Main component	Root	Root (%)	Accumulated %
CP1	9,78	32,62	32,62
CP2	6,75	22,52	55,15
CP3	3,47	11,58	66,74
CP4	2,94	9,81	76,55
CP5	1,66	5,54	82,10

CP6	1,39	4,64	86,74
CP7	0,92	3,07	89,82
CP8	0,77	2,59	92,42
CP9	0,48	1,60	94,02
CP10	0,36	1,20	95,22
CP11	0,28	0,93	96,16
CP12	0,24	0,80	96,97
CP13	0,20	0,69	97,66
CP14	0,18	0,622	98,28
CP15	0,15	0,51	98,80
CP16	0,12	0,41	99,22
CP17	0,08	0,27	99,50
CP18	0,074	0,24	99,74
CP19	0,028	0,096	99,84
CP20	0,018	0,060	99,90
CP21	0,012	0,041	99,94
CP22	0,009	0,030	99,97
CP23	0,003	0,012	99,99
CP24	0,0024	0,0080	99,99
CP25	0,000015	0,00005	99,99
CP26	0,0000045	0,000015	99,99
CP27	0,0000006	0,0000021	99,99
CP28	0,00	0,0000001	99,99
CP29	0,0000013	0,0000042	99,99
CP30	0,0000055	0,0000184	100,0

In genetic divergence studies between a group of genotypes, it is desirable for the accumulated variance in the first two principal components to exceed 80% (Cruz & Carneiro, 2003). However, from the results obtained, the accumulated variance between the components did not show a concentrated distribution in the first two components, which were responsible for explaining only 55.16 of the variation.

The first four components alone were able to explain 76.64% of the total variability between the accessions. It can also be seen that because the variance was diluted in the first four components, none of them, individually, was able to summarise a satisfactory amount of information.

These results are in line with those reported in the literature, such as those obtained by Choer and Silva (2000) who, using factor analysis, found that information from the four factors was needed to explain 76.60% of the variance between pumpkin accessions, as the first two concentrated only 53% of the variation. A similar result was reported by Martinello et al. (2003) and Modesta et al. (2005), who obtained only 57% and 58.5% of the variation in the first two components in okra and passion fruit, respectively.

Principal component analysis reduced the 10 descriptors in three ripening stages to just two, with the first (ascorbic acid content) being discrepant in all three stages and the second only in one stage. Components CP1, CP2 and CP3 explained 32.63, 22.52 and 11.59 per cent of the total variance of the accessions, respectively. The rest of the components accounted for only 13.3% of the total variance of the accessions.

Because the first four characteristics explained the most variation, it was necessary to obtain a three-dimensional graph (Figure 5), where the 25 accessions were separated into four distinct groups. Group I included accessions 1, 3, 9, 15, 23, 24 and 25; the second group was made up of accessions 4, 5, 7, 8, 11, 12, 13, 14, 16, 19 and 22; group III included accessions 2, 6, 10, 18, 20 and 21; the fourth group was isolated and consisted only of accession 17.

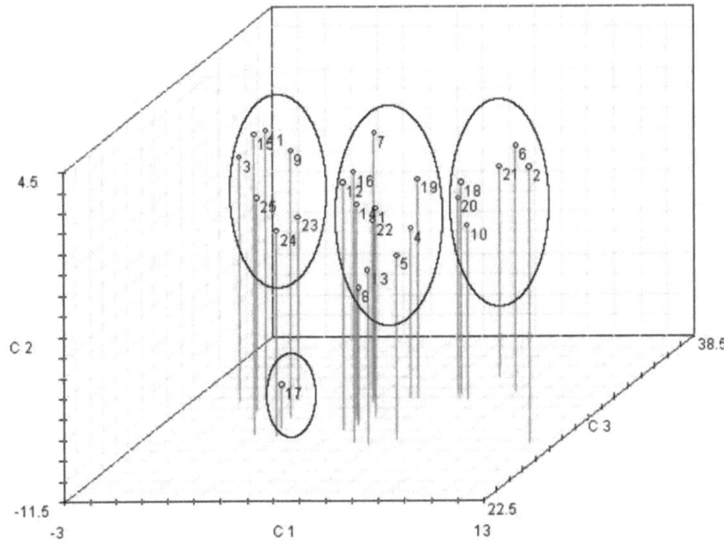

Figure 3 - Three-dimensional graph of the first four eigenvalues expressing the maximum total variation.

1 = ACE 001; 2 = ACE 002; 3 = ACE 008; 4 = ACE 009; 5 = ACE 010; 6 = ACE 011; 7 = ACE 012; 8= ACE 014; 9 = ACE 018; 10 = ACE 020; 11 = ACE 022; 12 = ACE 024; 13 = ACE 026; 14 = ACE 027; 15 = ACE 029; 16 = ACE 030; 17 = ACE 031; 18 = ACE 036; 19 = ACE 037; 20 = ACE 038; 21 = ACE 042; 22 = ACE 043; 23 = ACE 044; 24 = ACE 045; 25 = ACE 046.

The greatest genetic distance was observed between accessions 2 (ACE 002) and 17 (ACE 031), with 2.84 dissimilarity. On the other hand, the least genetically distant accessions were observed for accessions 12 (ACE 024) and 16 (ACE 030), with 0.65 dissimilarity.

Group I included the accessions with the highest levels of vitamin C, including: ACE 001, ACE 018, ACE 044 and ACE 046, which surpassed all the other accessions in the population, with values above 3000 mg per 100 g of pulp at the **'de vez'** stage of **ripeness. This** group also included accessions with high soluble solids, as can be seen in accessions ACE 001, with 9.90 °brix, and ACE 029, with 9.10 °Brix, at the ripe stage.

The second group had the highest number of accessions (44 per cent) of the 25 studied. However, group IV was represented by just one individual, accession ACE 031 (17). Using the hierarchical UPGMA method (Figure 06). It was found that there was agreement between the number of groups formed and the accessions thus grouped into them. Using a cut-off point of 1.3%, it was possible to see that ACE 002 and ACE 031 were isolated from the others. A group of six individuals was formed: the accessions: ACE 029, ACE 018, ACE 046, ACE 044, ACE 022 and ACE 008. The rest of the accessions were grouped together in the last group.

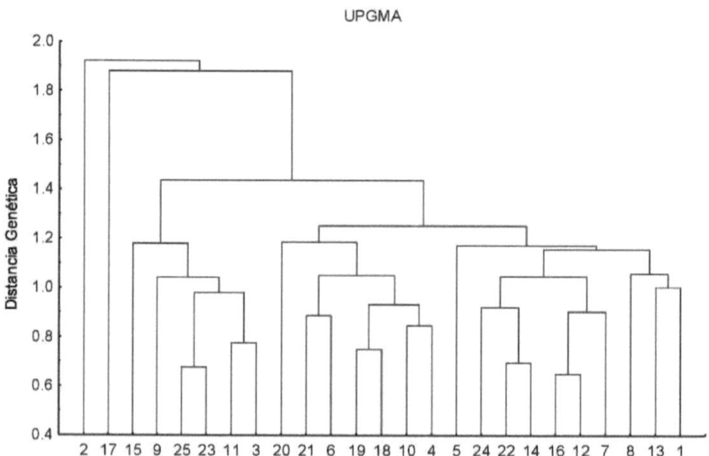

Figura 4 - Dendrogram representing the genetic divergence between the 25 acerola accessions, obtained by the UPGMA method, using the average Euclidean distance as a measure of dissimilarity.

4.6.1 Relative importance of characters for genetic diversity

The relative importance of the characters was assessed using Singh's method (1981), using 10 physicochemical characteristics at three stages of ripeness (Table 11). It was determined that only Vitamin C, for the three stages, contributed 98.84, while nine characteristics contributed less than 1%, therefore of little importance for the study of genetic diversity. The greatest contribution of this characteristic was seen in the semi-ripe stage (44.59%), followed by the ripe stage (27.57%) and the overripe stage (27.68%).

Table 11 - Relative contribution of 10 physico-chemical characters at three stages of ripeness to divergence in acerola, using Singh's method (1981).

Variable	S.j	Value in %
D1	2219,55	0,0006
D2	3638,21	0,0010
D3	2499,31	0,0007
C1	2205,00	0,0006
C2	2402,26	0,0007

C3	2151,12	0,0006
L1	83208,95	0,0231
L2	36316,77	0,0101
L3	16621,44	0,0046
A1	55019,04	0,0153
A2	32794,92	0,0091
A3	25558,35	0,0071
B1	43628,63	0,0121
B2	31983,49	0,0089
B3	17336,44	0,0048
H1	108648,76	0,0301
H2	70206,30	0,0195
H3	10853,72	0,0030
pH1	20,61	-
pH2	19,58	-
pH3	29,66	-
ATT1	25,38	-
ATT2	29,46	-
ATT3	40,34	-
SS1	208,49	0,0001
SS2	472,49	0,0001
SS3	651,61	0,0002
Vit 1	160774720,22	44,59
Vit 2	99412246,92	27,58
Vit 3	99795665,05	27,68

1 = 'once' stage; 2 = semi-mature stage; 3 = mature stage.
D= fruit diameter; C= fruit length; L, A, B H= Hunter parameters; pH= fruit hydrogen potential; ATT= total titratable acidity; TSS= total soluble solids; Vit c= Vitamin C content

4.7 RAPD test

Initially, DNA samples were used from four divergent accessions based on some morpho-agronomic characteristics (plant height, plant trunk diameter, skin colour of immature fruit and skin colour of mature fruit). Of the 72 primers tested, 25 provided clear products for amplification, as well as good repeatability. These were used throughout the population, generating a total of 92 polymorphic and 16 monomorphic marks (Table 12), where they were amplified with an average of 4.3 polymorphic bands per primer. The number of polymorphic marks varied from one to eight per primer, totalling 85.18% polymorphism. The polymorphic fragments used in the study can be considered sufficient for assessing genetic diversity in the species.

Table 12 - List of the 25 primers used from the OPERON Technologies series with their respective base sequences and number of polymorphic and monomorphic marks.

Initiator	Primer sequence	No. of polymorphic marks	No. of monomorphic marks	Total number of marks per primer
1-OPA 13	5'- CAGCACCCAC-3'	4	0	4
2-OPAB 01	5'- CCGTCGGTAG-3'	3	0	3
3-OPAB 09	5'- GGGCGACTAC-3'	3	1	4
4-OPAB 11	5'- GTGCGCAATG-3'	8	2	10
5-OPAB 13	5'- CCTACCGTGG-3'	4	0	4

6-OPAE 03	5'- CATAGAGCGG-3'	3	2	5
7-OPAE 14	5'- GAGAGGCTCC-3'	3	0	3
8-OPAE 16	5'- TCCGTGCTGA-3'	3	0	3
9-OPAE 17	5'- GGCAGGTTCA-3'	4	2	6
10-OPAF 13	5'- CCGAGGTGAC-3'	4	0	4
11-OPAF 20	5'- CTCCGCACAG -3'	3	1	4
12-OPAR 03	5'- GTGAGGCGCA-3'	3	0	3
13-OPAX 09	5'- GGAAGTCCTG-3'	6	0	6
14-OPAX 10	5'- CCAGGCTGAC-3'	2	0	2
15-OPAX 11	5'- TGATTGCGGG-3'	3	1	4
16-OPAX 14	5'- CACGGGCTTG-3'	6	0	6
17-OPAX 15	5'- CAGCAATCCC-3'	1	3	4
18-OPAX 16	5'- GTCTGTGCGG-3'	6	0	6
19-OPI 02	5'- GGAGGAGAGG-3'	3	1	4
20-OPI 03	5'- CAGAAGCCCA-3'	1	1	2
21-OPI 05	5'- TGTTCCAGGG-3'	4	0	4
22-OPI 06	5'- AAGGCGGCAG-3'	3	1	4
23-OPI 07	5'- CAGCGACAAG-3'	3	0	3
21-OPI 10	5'- ACAACGCGAG-3'	2	1	0
25-OPI 11	5'- ACATGCCGTG-3'	4	0	4
TOTAL		92	16	108

Salla et al. (2002), when evaluating the use of molecular markers to analyse genetic variability in acerola trees, found similar polymorphism results. The authors obtained around 90.8% polymorphism with four polymorphic bands per primer and stated that, although the crop has a narrow genetic base, the polymorphism generated by the markers is sufficient to characterise collections of acerole trees found in Brazil.

Emygdio et al. (2003), studying genetic divergence in common bean cultivars, found 85.6% polymorphism.

Studying genetic diversity in banana genotypes, Souza (2006) found that the number of total marks varied between one and four per primer, with 79.79% of the marks showing polymorphism.

Guimarães et al. (2007), in a morphological and molecular characterisation study of 22 fava bean accessions, obtained 76 bands, of which 60 were polymorphic and 16 monomorphic, corresponding to 78.94% polymorphism.

In a study to characterise and identify pear cultivars, Sawazaki et al. (2002) used 26 primers on 36 pear accessions and found 70.2% polymorphism.

Martinello et al. (2003), working with okra accessions, used 31 primers in 43 genotypes, which provided high polymorphism among the genotypes evaluated, amplifying a total of 103 polymorphic fragments, at a ratio of 3.3 marks per primer.

The nature of RAPD polymorphism depends on the types of primers used and their amplification products. The greater the amplification product, the greater the chances of detecting polymorphism and, consequently, obtaining more reliable results (Araújo et al., 2003). In the literature, studies related to the characterisation of acerole trees using genetic markers are scarce. It is therefore important to mention that among the primers used, some showed potential for use in diversity studies for the acerola crop, using RAPD markers, as they showed between six and eight polymorphic marks (Table 2). These include the primers OPAB-11, OPAX-09, OPAX-14 and OPAX-16.

Viana et al. (2003), studying the analysis of RAPD markers in passion fruit, showed that of the 100 primers tested, 14 showed the greatest potential for use: OPAC-15, OPAC-20, OPAD-01, OPAD-11, OPAD-14, OPAD-16, OPAD-19, OPAE-01, OPAE-08, OPAE-09, OPAE-10, OPAE-14, OPAE-18 and OPAE-19, which showed at least five or more polymorphic bands.

4.7.1 Genetic Diversity

Diversity was found among the 48 accessions studied using the RAPD marker technique, indicating that this technique was effective in identifying diversity in the acerola population. The amplification product was used to calculate the genetic similarity between the accessions. Figure 5 shows the electrophoretic pattern obtained with the OPAB-11 primer.

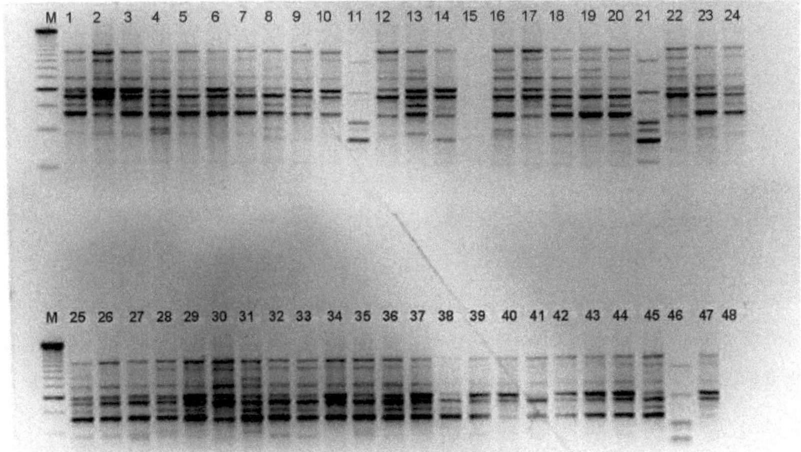

Figure 5 - RAPD polymorphic bands of the 48 accessions of the Pesagro-RIO acerole population

The complement of the Jaccard coefficient provided an estimate of the genetic dissimilarity between the accessions evaluated. The dissimilarities between the accessions ranged from 0.180 to 0.588, with an average genetic distance of 0.361. The minimum dissimilarities were recorded between accessions ACE 041 and ACE 043. On the other hand, the greatest dissimilarities were recorded between accessions ACE 023 and ACE 033 and between ACE 001 and ACE 017, with a genetic distance of 0.588 and 0.406, respectively.

In this work, there was no previous knowledge of the genetic variability of the parents of this population, however, the level of diversity obtained in this study suggests that it has high genetic variability. It is likely that the high degree of polymorphism found in the 48 genotypes is related to the fact that this species has predominantly cross-fertilised plant characteristics. Lopes et al. (2002), studying the isoenzyme polymorphism of acerola, suggest that the high degree of polymorphism observed in this crop is indicative of a considerable rate of crossing, confirming the predominance of allogamy in acerola. Additionally, Salla et al. (2002) evaluated the genetic variability between 24 acerola accessions using RAPD markers and reported that the genetic variability obtained by the markers may be related to the intense use of seeds in seedling production.

Diversity studies suggest that there is a tendency for the germplasm of tree and shrub plants, both allogamous and autogamous, with a high rate of allogamy, to show high polymorphism (Oliveira et al., 2007), especially in those that are poorly improved. There is therefore the possibility of obtaining significant genetic gains by using some of these accessions in future breeding programmes.

Based on the results presented here, in addition to the molecular characterisation of the accessions, this study has provided the possibility of future recommendations of genotypes to be used by acerola producers, based on genetic divergence associated with studies of morphological characteristics such as ascorbic acid content.

4.7.2 Cluster analysis using the Tocher and UPGMA methods

The Tocher optimisation method separated the 48 accessions into 13 distinct groups with 35 individuals (72.92%) in the first group (Table 13). Group I was thus made up of: ACE 041, ACE 043, ACE 005, ACE 040, ACE 003, ACE 044, ACE 037, ACE 045, ACE 006, ACE 010, ACE 002, ACE 032, ACE 015, ACE 012, ACE 036, ACE 038, ACE 035, ACE 013, ACE 004, ACE 047, ACE 042, ACE 046, ACE 034, ACE 008, ACE 011, ACE 022, ACE 027, ACE 019, ACE 014, ACE 016, ACE 021, ACE 039, ACE 007, ACE 031 and ACE 48. Generally, the groups formed by a large number of accessions group together the pairs with the smallest distances, since the size of the group is delimited by an average distance between the pairs of individuals.

With regard to the accessions that make up group I, similarities were observed between individuals in terms of chemical and morphological characteristics based on the preliminary descriptors indicated for acerola.

Salla et al. (2002), comparing the data obtained by RAPD markers with chemical and morphological characteristics, observed that group I associated eight accessions. These included the accessions UEL-26 and Iapar-1, whose main characteristic was similar vitamin C concentrations. The authors used data obtained by Carpentiere-Pípolo (2000) for the comparisons.

Group II was made up of accessions ACE 009 and ACE 033. The genetic distance observed between the accessions was 0.303.

The other groups (III, VI, VII, VIII, IX, X, XI, XII and XIII) only had one access each, ACE 025, ACE 026, ACE 020, ACE 018, ACE 029, ACE 001, ACE 028, ACE 017, ACE 030, ACE 024 and ACE 023, respectively.

Table 13 - Grouping of the 48 accessions from the Pesagro-RIO acerole population by the Tocher method, using the Arithmetic Complement of the Jaccard Index in RAPD markers.

Groups	Access
I	ACE 041, ACE 043, ACE 005, ACE 040, ACE 003, ACE 044, ACE 037, ACE 045, ACE 006, ACE 010, ACE 002, ACE 032, ACE 015, ACE 012, ACE 036, ACE 038, ACE 035, ACE 013, ACE 004, ACE 047, ACE 042, ACE 046, ACE 034, ACE 008, ACE 011, ACE 022, ACE 027, ACE 019, ACE 014, ACE 016, ACE 021, ACE 039, ACE 007, ACE 031 and ACE 48
II	ACE 009 and ACE 033
III	ACE 025
IV	ACE 026
V	ACE 020
VI	ACE 018
VII	ACE 029
VIII	ACE 001
IX	ACE 028
X	ACE 017
XI	ACE 030
XII	ACE 024
XIII	ACE 023

Because the first group had 35 accessions, which represents 72.92% of the accessions in the entire population, a new separation was carried out using the Tocher method, using only the individuals in group I. The subgrouping is described in Table 14, which shows the formation of 14 groups.

The results of the grouping showed that subgroup I brought together 17 accessions, equivalent to 48.57 per cent of the 35 accessions that belonged to group I and were regrouped. The accessions ACE 004, ACE 005, ACE 010, ACE 002, ACE 011, ACE 044, ACE 040, ACE 013, ACE 043, ACE 003, ACE 045, ACE 007, ACE 039, ACE 038, ACE 041, ACE 036 and ACE 008 remained in the same group in both the first and second groupings.

Of the new groups obtained (Table 14), five added two accessions each. These are: group II, with accessions ACE 032 and ACE 047; group III, with accessions ACE 014 and ACE 019; group IV, with accessions ACE 042 and ACE 046; group V, with accessions ACE 031 and ACE 034 and group VI, with accessions ACE 027 and ACE 035.

New groups were also formed with isolated accessions, such as groups VII, VIII, IX, X, XI, XII, XIII and XIV, made up of accessions ACE 048, ACE 022, ACE 012, ACE 006, ACE 016, ACE

021, ACE 015 and ACE 037, respectively.

The most divergent accessions were observed in isolated groups. According to Vieira et al. (2005), groups formed by just one individual suggest that these individuals are more divergent than the others.

Costa (2004), studying genetic diversity between accessions of Capsicum SSP. based on RAPD markers, used the Tocher optimisation method and obtained the formation of four large groups, with the first and second being divided twice into smaller groups. For this work, group I was subdivided into two smaller groups, bringing together several similar accessions (C. annuum, C. chinese and C. frutescens) and group II, which in turn was also subdivided twice, bringing together accessions of greater similarity.

Table 14 - Subgrouping of the accessions that make up group I (Table 13), by the Tocher method, using the Arithmetic Complement of the Jaccard Index in RAPD markers.

Groups	Access
I	ACE 004, ACE 005, ACE 010, ACE 002, ACE 011, ACE 044, ACE 040. ACE 013, ACE 043, ACE 003, ACE 045, ACE 007, ACE 039, ACE 038. ACE 041, ACE 036, ACE 008.
II	ACE 032 and ACE 047
III	ACE 014 and ACE 019
IV	ACE 042 and ACE 046
V	ACE 031 and ACE 034
VI	ACE 027 and ACE 035
VII	ACE 048
VIII	ACE 022
IX	ACE 012
X	ACE 006
XI	ACE 016
XII	ACE 021
XIII	ACE 015
XIV	ACE 037

Based on the data generated by the dissimilarity matrix, the dendrogram with the clusters shown in Figure 6 was obtained using the UPGMA clustering method, where the Y axis shows the percentage distances between the accessions and the X axis shows the 48 accessions.

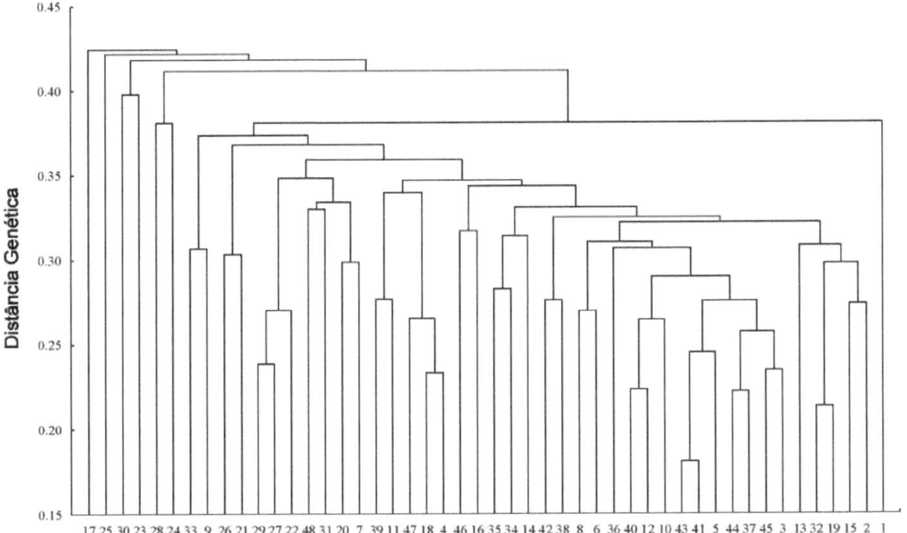

Figure 6 - Dendrogram representing the genetic divergence between the 48 acerola accessions, obtained by the UPGMA method, using the arithmetic complement of the Jaccard index, based on RAPD markers.

Fourteen groups were delimited, considering dissimilarity relative to 34% of the delimitation point in the dendrogram. Group I included the ACE 001 access. Group II is made up of 23 individuals, including the following accessions: ACE 035, ACE 034, ACE 014, ACE 042, ACE 038, ACE 008, ACE 006, ACE 036, ACE 012, ACE 040, ACE 010, ACE 0043, ACE 041, ACE 005, ACE 044, ACE 037, ACE 045, ACE 003, ACE 013, ACE 032, ACE 019, ACE 015, and ACE 002. Group III is made up of accesses: ACE 016 and ACE 046. Group IV is made up of the accessions: ACE 004, ACE 018, ACE 047, ACE 011 and ACE 039. group V - accesses: ACE 007, ACE 020, ACE 048 and ACE 031; group VI - accesses ACE 027, ACE 029 and ACE 022; group VII - accesses ACE 021 and ACE 026 ; group VIII- accesses ACE 009 and ACE 033; group IX- access ACE 024; group X- access ACE 028; group XI- access ACE 023; group XII- ACE 030; group XIII- ACE 025 and group XIV- ACE 017. The results showed that, for the two grouping methods used, it was possible to observe partial agreement between the results, since some accessions did not remain in the same group. On the other hand, some groups were formed by isolated genotypes, which were the same in both grouping methods.

4.8 Concordance between RAPD markers and morpho-agronomic characteristics.

Evaluations were made based on the list of minimum descriptors for acerola (Oliveira et al., 1998), and the following characteristics were chosen for comparison between the accessions grouped together: crown conformation, crown branching, general shape of the corolla edges, skin colour of the immature fruit, skin colour of the ripe fruit and size of the ripe fruit, which are important in the morphological characterisation of acerola and the main reference parameter for this crop, i.e. vitamin C content (Table 15). Due to the lack of initial fruiting of the materials, only 38 accessions with the characteristics studied were tabulated. On the other hand, only 25 accessions had enough fruit to analyse the ascorbic acid content.

Table 15 - Morpho-agronomic descriptors and ascorbic acid content for the acerola accessions from the Pesagro-Rio population.

Access	CC	RC	FGBF	CCFI	CCFM	TMF	Vit C (mg/100g)
ACE 001	Erect	very	wavy	green	purple	medium	2158
ACE 002	Erect	little	wavy	green	purplish	small	1444
ACE 003	-	-	-	-	-	-	-
ACE 004	Globular	little		purplish	purplish	medium	-
ACE 005	Globular	very	lightly	green	purplish	large	-
ACE 006	-	-	-	-	-	-	-
ACE 007	Erect	very		green	purplish	medium	-
ACE 008	Globular	very		green	purplish	large	1116
ACE 009	Intermediate	very	wavy	green	red	large	1935
ACE 010	Erect	very	straight	green	red	medium	1132
ACE 011	Erect	very	lightly	green	red	medium	1273
ACE 012	Erect	very	wavy	green	red	medium	1667
ACE 013	-	-	-	-	-	-	-
ACE 014	Erect	very	wavy	green	red	medium	2218
ACE 015	-	-	-	-	-	-	-
ACE 016	Intermediate	very	wavy	green	purple	large	-
ACE 017	Globular	very	wavy	green	purplish	large	-
ACE 018	Intermediate	very	lightly	green	purplish	large	1429
ACE 019	-	-	-	-	-	-	-
ACE 020	Erect	very	wavy	green	purplish	medium	1890
ACE 021	Globular	very	wavy	purplish	red	medium	-
ACE 022	Globular	very	wavy	purplish	purplish	large	1667
ACE 023	-	-	-	-	-	-	-
ACE 024	Intermediate	very	lightly	green	purple	large	1459
ACE 025	Globular	very	wavy	green	red	medium	-
ACE 026	Erect	very	wavy	green	red	medium	1727
ACE 027	Intermediate	very	wavy	purplish	purplish	large	1280
ACE 028	Globular	very	wavy	green	red	large	-
ACE 029	Globular	very	wavy	green	red	medium	1280
ACE 030	Intermediate	very	wavy	green	purplish	large	1429
ACE 031	Globular	very	wavy	green	red	large	2575
ACE 032	-	-	-	-	-	-	-
ACE 033	-	-	-	-	-	-	-
ACE 034	Erect	very	wavy	green	red	medium	-
ACE 035	Erect	very	wavy	green	red	medium	-
ACE 036	Erect	very	wavy	green	red	large	1637
ACE 037	Erect	very	wavy	green	red	large	1459
ACE 038	Erect	very	wavy	purplish	purplish	medium	2024
ACE 039	Erect	very	wavy	green	red	medium	-
ACE 040	-	-	-	-	-	-	-
ACE 041	-	-	-	-	-	-	-
ACE 042	Intermediate	very	wavy	green	red	medium	1980

ACE 043	Intermediate	very	wavy	green	red	medium	2024
ACE 044	Globular	very	wavy	purplish	red	large	1771
ACE 045	Globular	very	wavy	purplish	purplish	large	1161
ACE 046	Erect	very	wavy	purplish	purplish	large	2470
ACE 047	Intermediate	very	wavy	green	purplish	medium	-
ACE 048	Erect	very	wavy	purplish	red	medium	-

CC = crown conformation (1-globular; 2-intermediate; 3-Erect); RC = crown branching (1- Slightly branched; 2- branched; 3-very branched); FGBF = general shape of the mature leaf edges (1- straight edge; 2- slightly wavy edge; 3- wavy edge); CCFI = colour of the immature fruit skin (1- green; 2- yellowish green; 3- purplish green; 4- greenish purple; 5- purple); CCFM = ripe fruit skin colour (1- yellow; 2- orange; 3- pink; 4- red; 5- purplish red; 6- purple); TFM = ripe fruit size (1- small fruit (12-16 mm); 2- medium fruit (17-23mm); 3- large fruit (24-30mm) and Vit C = ascorbic acid content (mg/100g).

Group I was represented only by accession ACE 001, which stood out from the other accessions in its ascorbic acid content (2158 mg per 100g of pulp), which was the fourth highest of all the others. In addition to this variable, this access had a very branched canopy, the colour of the skin of the immature fruit was green, the colour of the peel was purplish and the fruit was physically medium-sized (Table 15).

There was similarity between the groups of accessions for the descriptors assessed (Table 15). Group II comprised 47.92 per cent of the accessions, with 23 individuals: ACE 035, ACE 034, ACE 014, ACE 042, ACE 038, ACE 008, ACE 006, ACE 036, ACE 012, ACE 040, ACE 010, ACE 0043, ACE 041, ACE 005, ACE 044, ACE 037, ACE 045, ACE 003, ACE 013, ACE 032, ACE 019, ACE 015, and ACE 002. In this group, 56.25 of the accessions had an erect crown and 93.75% of the individuals had a highly branched crown and leaves with a wavy edge, except for ACE 002, which had a slightly branched crown and ACE 010, which had leaves with a slightly wavy edge. With regard to the colour of the skin of the immature fruit, 75% of the accessions were green and only 10% (four individuals) had a purplish green colour. These results were discordant when compared with Pípolo et al. (2002), who described three acerola cultivars using morpho-agronomic descriptors and observed 100% similarity between the descriptors studied for all the cultivars selected.

The ascorbic acid content of the ripe fruit ranged from 1324 mg per 100g of pulp in access ACE 010 to 2575 mg per 100g of pulp in access ACE 014. These values were much higher than those found by Brunini et al. (2004), who recorded ascorbic acid contents ranging from 243.8 to 818.17 mg per 100 g of pulp in acerolas from Aparecida do Salto-SP, but similar to the values found by Lopes et al. (2000), who found accessions with ascorbic acid contents of up to 2246 mg/100 g of pulp. According to Bliska & Leite (1995), an ascorbic acid content above 1200 mg per 100g of pulp is considered a minimum benchmark for the product's acceptance on the foreign market. This therefore demonstrates the potential of the Pesagro-Rio acerola accessions, which meet the minimum standards for export.

Also, according to the morpho-agronomic variability between the accessions, three groups were formed comprising two accessions each (Table 11). Group III includes accessions ACE 016 and ACE 046, which have a highly branched crown, wavy corolla edges and large fruits. Accession ACE 046 had an ascorbic acid content of 2470 mg/100g of pulp. Group V, represented by accessions ACE 007 and ACE 020, showed an erect canopy and a very branched canopy, immature fruit with a green colour and a purplish red colour for the ripe fruit. Group VII, with accessions ACE 021 and ACE 026, had a very branched canopy, wavy corolla edges and medium-sized fruit. Group VIII, made up of accessions ACE 033 and ACE 009, had fruit with an ascorbic acid content of 1935 mg/100 g of pulp and was characterised by large fruit.

Group VI, made up of accessions ACE 027, ACE 029 and ACE 022, was characterised by its highly branched canopy. According to Pípolo et al. (2002), the acerola canopy selected for planting in a clonal orchard was of the intermediate type. This group includes the individual with the second highest ascorbic acid content of all those analysed (ACE 022 with 2619 mg/100g of pulp).

The other groups were made up of isolated accessions: group IX, represented by accession ACE 024; group X, by accession ACE 028; group XI, which included accession ACE 023; group XII, made up of accession ACE 030; group XIII, made up of accession ACE 025 and group XIV, represented by accession ACE 017, where the purplish red colour of the skin of the ripe fruit and the large size of the fruit were observed.

The results showed that the divergent accessions ACE 001, ACE 014, ACE 031, ACE 038, ACE 043 and ACE 046 (Table 1, Figure 1) stood out from the rest because they had a greater number of characteristics of agronomic interest, considered to be of greater commercial importance and indicated for vegetative propagation and clonal evaluation, as they had a high ascorbic acid content, reddish ripe fruit skin colour and fruit size ranging from medium to large.

Lopes et al. (2000), evaluating the physico-chemical characteristics of 112 acerola accessions, found several promising accessions that were recommended for clonal evaluation with a view to their use as varieties. The authors found fruit with ascorbic acid content ranging from 1761 mg to 2220 mg per 100 g of pulp and total soluble solids ranging from 5.8 to 10.1 ° Brix.

Pípolo et al. (2000), with the aim of identifying and selecting parental genotypes of acerola, based on multivariate genetic divergence in 14 genotypes, managed to divide them into three groups, indicating seven most promising crosses based on ascorbic acid content: AM Mole belonging to group III, with the genotypes PR AM, N° 18, PR 17, PR 16, Eclipse, AM 22 and Dominga, all belonging to group I. The authors also suggested that large fruit with a larger diameter result in the selection of fruit with a greater amount of pulp.

CHAPTER 5

SUMMARY AND CONCLUSIONS

The results obtained by analysing the multicategorical variables of the acerola accessions show that there is genetic variability for 70.5% of the characters studied.

Three quantitative characteristics considered to be of greatest relative importance for the genetic divergence study were selected: stem diameter, which showed the greatest relative contribution, followed by mature leaf length and maximum leaf width.

Fruits in the 'once' stage showed the highest values of ascorbic acid content, regardless of the access, with a gradual reduction until the final ripeness of the fruit, indicating that there is a clear correlation between ascorbic acid content and the stage of ripeness.

The RAPD marker technique was effective for studying diversity among the accessions, showing that the population has ample genetic variability.

We found primers with potential for use in diversity studies for the acerola crop using RAPD markers, as they showed between six and eight polymorphic marks, including the following primers: OPAX-09, OPAX-14, OPAX-16 and OPAB-11.

The Tocher Optimisation and UPGMA methods were in agreement and satisfactory for forming fairly homogeneous groups for the characteristics assessed, both for the RPAD markers and the morpho-agronomic characters.

There was an increase in pH and TSS between the accessions from the 'once' stage to the semi-mature stage.

The results showed that the accessions ACE 001, ACE 014, ACE 031, ACE 038, ACE 043 and ACE 046 could be suitable for vegetative propagation and clonal evaluation, as they had a high ascorbic acid content, reddish ripe fruit skin colour and large fruit size.

BIBLIOGRAPHICAL REFERENCES

Alves, R. E. & Menezes, J. B. (1995) Botany of the acerola tree. In: São José, A. R. & Alves, R. E. ed. Acerola in Brazil: Production and Market. Vitória da Conquista BA, UESB, p.7-14.

Alves, R. E., Menezes, J. B., Silva, S. M. (1995) Harvest and post-harvest of acerola. In: São José, A.R., Alves, R.E. Acerola no Brasil: produção e mercado. Vitória da Conquista: DFZ/UESB, p. 77-89.

Alves, R. E. (1996) Characteristics of fruit for export. In: Gorgatti Netto, A., Ardito, E. F. G., Garcia, E. E. (Eds.) Acerola for export: harvest and post-harvest procedures. Brasília: EMBRAPA-SPI. p.9-12. (FRUPEX Technical Publications Series, 21).

Amaral Júnior, A. T. (1994) Isoenzymatic multivariate analysis of genetic divergence between strawberry accessions (Cucurbita maxima Duchesne). 95 f. Dissertation (Master's Degree) - Federal University of Viçosa, Viçosa.

Andrade, J. M. B., Brandão Filho, J. V. T., Vasconcelos, M. A. S. (1995) Effect of pruning on the productivity of acerola (Malpighia glabra L.) in the first year. Revista Brasileira de Fruticultura, Cruz das Almas, v.17, n.2, p.45-49.

Association of Official Analytical Chemicals (1970) Official methods of analysis. Washington, 101. 5p.

Association of Official Analytical Chemical (1984) Official Methods of Analysis. Washington, 101. 5p.

Araújo, P. S. R., Minami, R. (1994) Acerola. Campinas: Cargil Foundation, 81p.

Araújo, E. S., Santos, A. M., Areias, R. G. B. M., Souza, S. R., Fernandes, M. S (2003). Use of RAPD to analyse genetic diversity in rice. Agronomia, V.37, N°1, P.33 - 37.

Asenjo, C. F. (1980) Acerola. In: Nagy, S., Shaw, P. E. Tropical and subtropical fruits: conposition, properties and uses. Westport. AVI. p. 341-74.

Assis, S. A., Lima, D. C., Oliveira, O. M. M. F. (2001) Activity of pectinmethylesterase, pectin content and vitamin C in acerola fruit at various stages of fruit development. Food Chemistry, v.74, p.133-137.

Batista, F. S., Muguet. B. R. R. Beltrão (1991) A. E. S. Behaviour and selection of aceroleira in Paraíba . In: Congresso Brasileiro de Fruticultura. 9. Fortaleza, 1989. Proceedings p. 26-32

Bento, C. S., Sudré, C. P., Rodrigues, R., Riva, E. M., Pereira, M. G. (2007) Qualitative and multicategorical descriptors for estimating phenotypic variability among pepper accessions. Scientia Agraria, v.8, n.2, p.149-156.

Blank, A. F., Carvalho Filho, J. L. S., Santos Neto, A. L., Alves, P. B., Arrigoni- Blank, M. F., Silva-Mann, R., Mendonça, M. C. (2004) Morphological and agronomic characterisation of basil and lavender accessions. Horticultura Brasileira, Brasília, v.22, n.1, p. 113-116, jan-mar 2004.

Bliska, F. M. M., Leite, R. S. S. F. (1995) Economic aspects and the market. In: São José, A. R. Alves, R. E. Acerola no Brasil: produção e mercado. Vitória da Conquista: UESB, 1985.45p

Bosco, J., Filho, A. S. P., Neto, M. B. (1994) Phenological characteristics of acerola plants, In: Congresso Brasileiro de Fruticultura, Salvador, BA, v.1, p.87.

Brunini, M. A., Macedo, N. B., Coelho, C. V., Siqueira, G. F. (2004) Physical and chemical characterisation of acerolas from different growing regions. Revista Brasileira de Fruticultura, v. 26, n. 3,p. 486-489.

Cavalcante, H. C., Bff, T., Dornelles, A. L.C., Oliveira, J. R. P. (1998) Cytogenetic analysis of eight genotypes of acerola (Malpphighia punicifolia L) In: Congresso Brasileiro de Fruticultura, 15, Poços de caldas, MG. Annals of thc congress

Carpentieri-Pípolo, V., Prete, C. E. C., Gonzalez, M. G. N. (2002) New acerola cultivars (Malpighia emarginata D.C.). UEL 3 (Dominga) - UEL 4 (Lígia) - UEL 5 (Natália). Revista Brasileira de Fruticultura, Jaboticabal, v.24, n.1, p.124-126.

Carvalho, W., Fonseca, M. E. N., Silva, H. R., Boiteux, L. S., Giordano, L. B. (2005) Indirect estimation of lycopene content in fruits of tomato genotypes by colourimetric analysis. Horticultura Brasileira, Brasília, v.232 n.3, p.819-825, jul-set

Carvalho, R. A. (2000) Economic analysis of acerola production in the municipality of Tomé-Açú, Pará. Belém: Embrapa Amazônia Oriental. 21p. (Document, 49)

Coelho, Y. S., Ritzinger, R., Oliveira, J. R. P. (2003) Proacerola: Programme for the Development of Acerola Culture in the State of Bahia. In: Annual Meeting of the Inter-American Society of Tropical Horticulture, 49, Fortaleza, Abstract. Fortaleza: Inter-American Society of Tropical Horticulture, 303p.

Choer, E., Silva, J. B. (2000) Evaluation of genetic diversity among curcubita ssp. accessions through multivariate analysis. Agropecuária de Clima Temperado, Pelotas. V.3 p. 213-219

Conceição, M. P. J. (1997) Thermal degradation kinetics of anthocyanins in acerola juice (Malpighia glabra L.). Viçosa, 59p. Master's dissertation - Federal University of Viçosa (UFV).

Costa, A. P. (2004) Evaluation of agronomic characteristics in varieties and molecular diversity in grapevine varieties, hybrids and species. Dissertation (Master's Degree) Universidade Estadual do Norte Fluminense Darcy Ribeiro - UENF. Campos dos Goytacases, RJ. 95 p.:il

Costa, F. R. (2004) Genetic diversity among Capsicum ssp accessions based on RAPD. Dissertation (Master's Degree) - State University of Northern Rio de Janeiro Darcy Ribeiro.

Cruz, C. D, Carneiro, P. C. S. (2003) Biometric models applied to genetic improvement. Viçosa: UFV, v.2.p 415.

Cruz, C. D. (2006) Genes programme: multivariate analysis and simulation. Viçosa: ed. UFV. 175p.:il

Cruz, C.D., Regazzi, A.J. (1997) Biometric models applied to genetic improvement. Viçosa: UFV, 390p.

Donadio, L. C., Nachtigal, J. C., Sacramento, C. K. (1998) Exotic fruits. Jaboticabal: Funep, 279p.

Doyle, J. J. & Doyle, J. L. (1997) Isolation of plant DNA from fresh tissue. Focus, v.12:13-15.

De Rosso, V. V., Mercadante, A. Z. (2005) Carotenoid composition of two Brazilian genotypes of acerola (Malpighia punicifolia L.) from two harvests. Food Research International, v. 38, n. 8-9, p. 1073-1077, 2005.

Emygdio, B. M, Antunes, I. F, Nedei, J. L, Choer, E. (2003) Genetic diversity in local and commercial bean cultivars based on RAPD markers. Pesquisa Agropecuária Brasileira, v. 38, p.1165-1171.

Ferreira, M. E., Grattapaglia, D. (1998) Introduction to the Use of Molecular Markers in Genetic Analysis . Brasília: Embrapa - Cenargem, 3rd edition, 220p.

Folegatti, M. I. S., Matsuura, F. C. A. U. (2003) Products. In: Ritzinger, R., Kobayashi, A. K., Oliveira, J.R.P (eds) A cultura da aceroleira. Cruz das Almas, BA: Embrapa Mandioca e Fruticultura, p. 164-184.

Fontes, P. S. F. (2002) Nitrogen fertilisation and evaluation of banana cultivars (musa spp.) in the north-west of the state of Rio de Janeiro. Thesis (Master's in Plant Production) - Campos dos Goytacazes - RJ, Universidade Estadual do Norte Fluminense Darcy Ribeiro, 64p.

França, V. C., Narain, N. (2003) Chemical characterisation of the fruits of three varieties of acerola (Malpighia emarginata D.C.). Ciência e Tecnologia de Alimentos, Campinas, v.23, n.2, p. 157-160.

Gomes, J. E. , Perecin, D., Martins, A. B. G. Almeida, E. J. (1998) Correlations and direct and indirect effects on the selection process in acerola cultivation. In: Congresso Brasileiro de Fruticultura, 15, Poços de Caldas, MG. Annals of the congress.

Gomes, J. E., Pavani, M. C. M. S, Perecin, D., Martins, A. B. G. (2001) Floral morphology and reproductive biology of acerola genotypes. In: scientia agrícola, v. 58, n.3, p. 519 - 523.

IBRAF - Brazilian Fruit Institute (1995). Fruit by fruit solutions: acerola. 59p.

Jolliffe, I. T. Discarding variables in a principal component analysis (1973). II. Real data. Appl. Stat., v.22, p.21-31.

Loarce, Y., Gallego, R., Ferrer, E. A. (1996) comparative analysis of the genetic relationship between rye cultivars using RFLP and RAPD markers. Euphytica, Wageningen, v. 88, p. 107-115.

Lima, V. L. A. G., Melo, E. A. M., Gerra, N. B. (2007) Correlation between Anthocyanin Content and Chromatic Characterisation of Pulps from Different Acerole Tree Genotypes. Braz. J. Food Technol., Campinas, v. 10, n. 1, p. 51-55, jan./mar.

Lima, V. L. A. G., Mélo, E. A., Maciel, M. I. S., Prazeres, F. G., Musser, R. S., Lima, D. E. S. (2005) Total phenolic and carotenoid contents in acerola genotypes harvested at three ripening stages. Food Chemistry, v. 90, n. 4, p. 565-568

Lima, V. L. A. G., Melo, E. A., Maciel, M. I. S., Lima, D. E. S. (2003) Evaluation of anthocyanin content in frozen acerola pulp from 12 different acerola trees (Malpighia emarginata D.C.). Ciência e Tecnologia de Alimentos, v. 23, n. 1, p. 101-103.

Lopez Camelo, A. F., Gómez, P. A (2004) Comparison of colour indexes for tomato ripening. Horticultura Brasileira, Brasília, v.22, p.534-537.

Lopes, R., Paiva, J. R. (2002) Aceroleira. In: Bruckner, C.H. Melhoramento de Frutiras Tropicais. Editora Universidade Federal de Viçosa/UFV. p. 63-99.

Lopes, R, Bruckner, C. H, Lopes, M. T. G. (2002). Estimation of acerola crossing rate based on isoenzyme data. Pesquisa Agropecuária Brasileira, v. 37, Brasília.

Lopes, R., Bruckner, C. H., Finger, F. L., Lopes, M. T. G. (2000) Evaluation of fruit characteristics of acerola accessions. Revista Cers, v.47, n.274, p.627-638.

Martinello, G. E., Leal, N. R., Amaral Júnior, A. T., Pereira, M. G., Daher, R. F. (2003) Genetic diversity in okra based on RAPD markers. Horticultura Brasileira, Brasília, v. 21, n. 1, p. 20-25.

Matsuura, F. C. A. U., Cardoso, R. L., Folegatti, M. I. S., Oliveira, J. R. P., Oliveira, J. A. B., Santos, D. B. (2003) Physico-chemical evaluations of different acerola (Malpighia punicifolia L.) genotypes. Revista Brasileira de Fruticultura, Jaboticabal v.23 , n.3, p.602-606.

Marty, G. N. And Pennock, W. (1965) Práticas agronômicas para el cultivo comercial de acerola en Puerto Rico. Revista de Agricultura de Puerto Rico, 52: 107-111.

Milach, S. C. K. (1998) Molecular markers in plants. Porto Alegre, 141p.

Moreira, J. A. N., Santos, J. W., Oliveira, S. R. M. (1994) Approaches and methodologies for germplasm evaluation. Campina Grande: Embrapa-CNPA, 115 p.

Modesta, R. C., Gonçalves, E. B., Rosenthal, A., Silva, A. L. S., Ferreira, J. C. S. (2005) Development of the passion fruit profile, Ciênc. Tecnol. Aliment., Campinas, 25(2): 345-352, Apr.-Jun. 2005

Moura, C. F. H., Alves, R. E., Figueiredo, R. W., Paiva, J. R. (2007) Physical and physicochemical evaluations of fruit from acerola clones (Malpighia emarginata D.C.) Revista Ciência Agronômica, v.38, n.1, p.52-57,

Nass, L. L. (2001) Utilisation of plant genetic resources in breeding. In: NASS, L. L. et al. (Ed.) Recursos genéticos e melhoramento. Rondonópolis, UFMT. 1183p.

Neto, G. L. Melhoramento genético da aceroleira. (1995b) In: São José, A. R., Alves, R. E. (ed.) Acerola no Brasil, produção e mercado. Vitória da Conquista % BA: UESB, 160p.

Nogueira, R. J. M. C. (1997) Physiological Expressions of the Acerole Tree (Malpighia emarginata

D. C.) under adverse conditions. São Carlos, 207p. Thesis (Doctorate) - University of São Carlos.

Nogueira, R. J. M. C., Moraes, J. A. P. V., Burity, H. A. (2002) Effect of fruit ripeness on the physico-chemical characteristics of acerola. Pesquisa Agropecuária Brasileira, Brasília, v.37, n.4, p.463-470.

Oliveira, L. S., Rabelo, C. R., Aguiar, R. A., Moura, C. F. H., Miranda, M. R. A. (2007) Evaluation of post-harvest quality and antioxidant enzyme activity in six acerola clones (Malphigia ermaginata D. C) In: II Simpósio Brasileiro de Pós-Colheita de Frutas, Hortaliças e Flores. V. 2. p.382, Viçosa-MG

Oliveira, M. S. P, Amorim, E. P, Santos, G. B, Ferreira, D. F. (2007) Genetic diversity among açaizeiro accessions based on RAPD markers. Ciênc. agrotec., Lavras, v. 31, n. 6, p. 1645-1653.

Oliveira, J. R. P., Soares Filho, W. S. (1998) Situation of the acerola crop in Brazil and Embrapa Mandioca e Fruticultura's actions in genetic resources and improvement. In: Simpósio de Recursos Genéticos e Melhoramento de Plantas Para o Nordeste do Brasil, Petrolina, Anais... Petrolina: Embrapa SemiÁrido.

Oliveira, J. R. P., Soares Filho, W. S., Cunha, R. B. (1998) Guide to acerola descriptors: preliminary version. Cruz das Almas, BA: Embrapa - CNPMF, 22 p. (Document, 84).

Paiva, J. R., Alves, R. E., Santos, F. J. S. S., Barros, L. M., Almeida, A. S., Moura, C. F. H. M., Bezerra C. J., Pereira N. N. (2003) Preliminary Selection of Aceroleira Clones in the State of Ceará Ciênc. agrotec., Lavras. V.27, n.5, p.1038-1044.

Paiva, J. R., Alves, R. E., Correa, M. P. F., Freire, F. C. O., Braga Sobrinho, R. (1999a) Mass selection of acerola in commercial planting. Pesquisa Agropecuária Brasileira, Brasília, v.34, n.3, p.505-511.

Paiva, J.R., Alves, R.E., Santos, F.J.S. (2002) Performance of acerola clones in the State of Ceará. In: Congresso Brasileiro de Fruticultura, 17, Belém, Anais... Belém: SBF, CD-ROM.

Pípolo, V.C., Destro, D., Prete, C.E.C., Gonzales, M.G.N., Popper, I., Zanatta, S., Silva, F.A.M. (2000) Selection of acerola parental genotypes based on multivariate genetic divergence. Pesquisa Agropecuária Brasileira, v.35, n.8, p.1613-1619.

Rosa, M. S., Santos, P. P., Veasey, E. A.(2006) Interpopulation Agromorphological Characterisation in Oryza Glumaepatula. Bragantia, Campinas, V.65, N.1, P.1- 10.

Ruas, P. M., Ruas, C. F., Fairbands, J. D., Andersen, R. W., Cabral, J. R. S. (1995) Genetic relatonship among four varieties of pineapple, Ananas comosus, revealed by random amplified polymorphic DNA (RAPD) analysis. Brazilian Journal of Genetics, Ribeirão Preto, v. 18, n. 3, p. 413 - 416.

Salla, M. F. S., Ruas, C. F., Ruas, P. M., Pípolo, V. C. (2002) Use of molecular markers to analyse genetic variability in acerola (Malpighia emarginata D.C.). Revista Brasileira de Fruticultura, Jaboticabal v.24, n.1, p.015- 022.

Sawazaki, H. E., Wilson Barbosa, W., Colombo, C. A. (2002) Characterisation and identification of pear cultivars and selections using RAPD markers Rev. Brasileira de Fruticultura, Jaboticabal - SP, v. 24, n. 2, p. 447-452.

Simão, S. (1971) Manual de fruticultura. São Paulo: Agronômica Ceres, 530p.

Shimoya, A., Pereira, A. V., Ferreira, R. P., Cruz, C. D., Carneiro, P. C. S. (2002) Repeatability of forage characteristics of elephant grass. Scientia Agricola, 59:227-234.

Soares Filho, W. S., Oliveira, J. R. P. (2003) Introduction. In: Ritzinger, R., Kobayashi, A. K., Oliveira, J. R. P (eds) A cultura da aceroleira. Cruz das Almas, BA: Embrapa Mandioca e Fruticultura, p. 15-16.

Souza, C. M. P., (2006) DNA **Fingerprinting via RAPD markers and evaluation of** genetic divergence in banana genotypes (**Musa** spp.) Thesis (Master's in Plant Production) - Campos dos Goytacazes - RJ, Universidade Estadual do Norte Fluminense Darcy Ribeiro

Sudré, C. P., Cruz, C. D., Rodrigues, R., Riva, E. M., Júnior, A. T. A., Silva, D. J. H. Da, Pereira, T. N. S. (2006.) Multicategorical variables in determining genetic divergence between pepper and paprika accessions. Horticultura Brasileira, Brasília, v. 24, n. 1, p. 88-93, jan./mar.

Sudré, C. P, Rodrigues, R., Riva, E. M., Karasawa, M., Amaral Júnior, A. T. (2005) Genetic divergence between pepper and paprika accessions using multivariate techniques. Horticultura Brasileira, v.23, n1, 22-27p.

Ulanovsky, S. (2002) Use of molecular markers in detection of synonymies and homonymies in grapevines (Vitis vinifera L.). Scientia Horticulturae, Amsterdam, v.92, p.241-254.

Viana, A. P., Pereira, T. N. S., Pereira, M. G., Amaral Jr, A. T., Souza, M. M., Maldonado, J. F. M. (2003) Genetic diversity among commercial genotypes of yellow passion fruit (P. edulis f. flavicarpa) and among native Passifloras species determined by RAPD markers. Revista Brasileira de Fruticultura, Jaboticabal v. 25, n. 3, p.489-493.

Vieira, E. A., Fialho, J. De F., Faleiro, F. G., Fukuda W. M. G., Junqueira N. T. V. (2005) Genetic variability for morphological characters among accessions of the

Embrapa Cerrados cassava germplasm bank. In: CONGRESSO BRASILEIRO DE MANDIOCA, 11, 2005b, Campo Grande. Proceedings... Campo Grande, MS. 1CD-ROM.

Yamashita, F., Benassi, M.T., Tonzar, A.C. (2003) Acerola products: vitamin C stability studies. Ciência e Tecnologia de Alimentos, Campinas, v.23, n.1, p.92-94.

Wadt, L.H.O. (1997) Evaluation of genetic divergence in coconut palm (cocos nucifera L.) using markers in composite samples of individual or composite plants. 1997, UENF, Master's thesis

Wolf, A. B., Macrae, E. A., Spooner, K. l., Redewell, R. J. (1997). Changes to physical properties of the cell wall and polyuronides in response to heat treatment of' Fuyu ' persimmon that alleviate cilling injury. Journal of American Society for Horticultural Sciense, Alexandria, v. 122, p. 698-702.

Printed by Books on Demand GmbH, Norderstedt / Germany